Digital Timi
Circuits

J. B. Gosling

Senior Lecturer
Department of Computer Science, Manchester University

Edward Arnold

First published in Great Britain 1985 by
Edward Arnold (Publishers) Ltd,
41 Bedford Square,
London WC1B 3DQ

Edward Arnold,
300 North Charles Street,
Baltimore, Maryland, USA

Edward Arnold (Australia) Pty Ltd,
80 Waverley Road, Caulfield East,
Victoria 3145, Australia

Gosling, John B.
Digital timing circuits.
1. Digital electronics 2. Electronic
circuit design
I. Title
621.3815'3 TK 7868.D5

ISBN 0-7131-3524-7

Typeset in Hong Kong by
Asco Trade Typesetting Ltd.

Preface

One of the most difficult parts of the design of any system, be it digital or analogue, on printed circuit boards or silicon, is the timing. Yet the subject receives scant attention in the literature. The author has felt for some time that there was a need for a good text, bringing together all the bits and pieces scattered over a chapter here and a paragraph there. This book is an attempt to do just that.

In preparing the work the greatest difficulty has been to decide the background of potential readers. A minimum background is assumed, but this, no doubt, will be too much for some, and too little for others. It is assumed that readers are familiar with logic gates, but not with bistable, monostable etc. circuits. The book by Gibson (see Appendix A) is recommended as a good text describing this work, as well as of the design and use of flip-flops. A certain amount of mathematical knowledge is also assumed, in particular the ability to perform integration, and use complex numbers. On the electrical side, some knowledge of circuit techniques is necessary. Firstly, a basic knowledge of electrical components—resistance, capacitance, voltage sources, etc.—and the solution of simple networks using Kirchoff's laws. Secondly, the knowledge of the characteristics of a diode, and the use of the bipolar transistor. If this latter is not known, then the reader may be able to appreciate the circuit by simply regarding the device as a switch. The idea of saturation of a bipolar transistor, and the use of Schottky diodes to prevent it, is explained, but only very briefly. Thirdly, a familiarity with the notion of input and output impedance of a device is presumed, although the intelligent reader should have little difficulty picking this up. Finally, a few results are derived by making use of 'sine wave' techniques, in particular making use of the $j\omega$ technique. Knowledge of this is not essential to the understanding of the book, merely to understand how certain conclusions are reached.

The text is unique so far as is known in the way it brings together the topics. It includes a discussion of the synchronisation problem, and a digital approach to transmission line theory which minimises the necessary mathematics and gives results quite adequate for the majority of real situations. The variety of circuits for pulse and clock waveform generation is also unique for a single text.

No claims to completeness are made. At least one important circuit is omitted, simply because of its complexity. In the area of transmission line techniques much has been omitted, largely for brevity. In particular, the graphical analysis methods of handling mismatched lines have been omitted.

Many people have contributed to the author's understanding of the subject. Mention is made particularly of Professor D B G Edwards of the Department of Computer Science in the University of Manchester, and of Professor D J Kinniment, now of the University of Newcastle upon Tyne. Comments from Dr A E Knowles and the Publisher's referee were extremely helpful, leading to a number of revisions and additions that have added much to the work.

J B Gosling

Manchester
1984

Acknowledgements

The following organisations have given permission for parts of their works to be used.

Mullard Ltd., *Signetics Technical Handbook* (1977). Figure 9-37 and parts of Fig. 9-19, relating to voltage controlled oscillators, are similar to Figs 6.10 and 6.11 of this text.

SGS (UK) Ltd., *Applications of Linear Microcircuits*, Vol. 1 (1969). Figures 3-48 and 3-51 relate to Figs 5.9 and 6.7 of this text.

Addison-Wesley Publishers Ltd.. Figure 9.14 of this text is modified from Fig. 7.7 of Mead and Conway, *Introduction to VLSI Systems* (1980).

Their permission is gratefully acknowledged.

Acknowledgement is also made to the University of Manchester for the examination questions used as examples. These are designated 'U of M', meaning 'University of Manchester, Department of Computer Science, 2nd Examination'.

Abbreviations

The reader should be aware of the following standard abbreviations:

M	mega	10^6
k	kilo	10^3
m	milli	10^{-3}
μ	micro	10^{-6}
n	nano	10^{-9}
p	pico	10^{-12}

Contents

1 The importance of timing in system design

Introduction

Time is a fundamental component of all dynamic systems, electrical or otherwise. Consider, for example, a motor car engine. In a family car there are usually four cylinders in which pistons move up and down. Every second time a piston reaches the top of the cylinder, a spark causes a mixture of petrol and air to explode. The spark must occur at just the right moment, or the piston may be forced backwards, against the effort of the other cylinders. The spark occurs only in alternate cycles. Within the four-stroke cycle, petrol and air must be admitted to the cylinder at a particular time, and combustion products exhausted at another. The whole sequence must be carefully **timed** if the best operation is to be obtained. The timing of all four cylinders is the same in terms of what happens, but each is staggered *in time* relative to the other three. This relative timing must be held within closely controlled limits.

For an electronic example, consider the arithmetic unit of a computer. This is capable of executing several different instructions (add, multiply, divide, etc.). The computing system provides a sequence of instructions to be performed one after the other. An add will generally take less *time* than a multiply. However, care must be taken to ensure that one instruction does not begin before the previous one has finished.

Figure 1.1 shows a possible structure for such a system. The following definitions will be useful in explaining this diagram, as well as for later use.

Definitions

A latch and a flip-flop can be regarded as 1-bit stores. They are **set** to a one or **reset** to a zero at any time by a signal applied to an appropriate input, and hence will hold in one of two states. They are examples of **two-state devices**. With most two-state devices it is also possible to set the state to a one or a zero from a 'data' input, but only at fixed times. The signal defining the fixed time comprises a regular series of pulses and is called a **clock**, or sometimes a **strobe**.

The **latch** has an output which follows the data input while the clock is in one state, but is unaffected by the data input during the other state of the clock. With the **flip-flop**, the data is transferred to the output following one of the *edges* of the clock signal. For a positive-going edge triggered flip-flop (to give it its full name), a clock edge going from a relatively low voltage to a relatively high voltage causes the output to change if such a change is required.*

Individual flip-flops and latches often have two outputs, which are in opposite logical states when the device is being operated properly (although in many cases it is possible to get both outputs to a one). One output is called the **true** output, or simply

*The terms 'latch' and 'flip-flop' are today generally used as described above. However, the term 'flip-flop' is sometimes used for both types of device, in which case they are distinguished by the terms 'level triggered' and 'edge triggered' for the latch and flip-flop respectively.

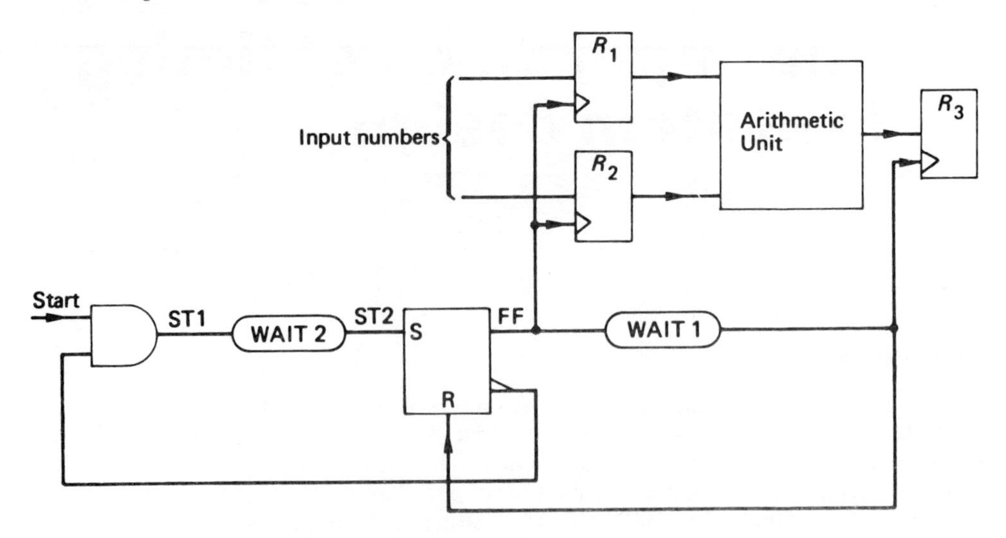

Fig. 1.1 Simplified arithmetic unit and control

the output. The other is the **not** or **inverse** output.

A **register** is a collection of flip-flops operated as a single unit, with a clock signal common to all bits. It often stores a 'word' of information in computer terms. Usually only the true output is available, due to pin limitations on the IC packages.

The terms 'set' and 'reset' as applied to a flip-flop imply driving the true output to a one or a zero respectively. This is sometimes done via special inputs allowing action independent of the clock. The term **load** as applied to a register or to a flip-flop means transferring information from the data inputs to the outputs at the appropriate clock time.

Figure 1.1 shows an arithmetic unit with rudimentary control. Two registers R_1 and R_2 hold the numbers to be added (say). When the two numbers are ready, the start signal sets a flip-flop, FF. This signal clocks R_1 and R_2, causing them to be loaded. Some time later (after WAIT1) the delayed signal causes R_3 to be loaded with the result. During this time the start pulse must be locked out to prevent an attempt to begin another operation, otherwise R_1 and R_2 may be altered before the end of a long instruction such as a multiply. Such an attempt will be handled by other circuits external to the unit shown. The NOT side of FF can be used to lock out the start signal.

Figure 1.2 shows details of the **timing**. The pointers run from a cause to an effect, thus illustrating how the state (or change of state) of one of the components affects another component. This is essentially a graph of logic levels against time; the time increases from left to right. The **waveforms** are shown such that a vertical line drawn through all signals represents the same time on all signals. The vertical axis is, in this case, the logical state, but it may represent voltage or current or other variable. Axes are not generally drawn, but values of variables (time, voltage, etc.) must be. If not otherwise stated, logic waveforms presume that a 'one' is represented by the higher of the two levels.

It is not necessary at present to understand this example completely. It *is* necessary to understand the major importance of timing.

The situation is, in fact, more complex than is shown here. Firstly, different instructions require different values of WAIT1. Secondly, real gates do not operate in

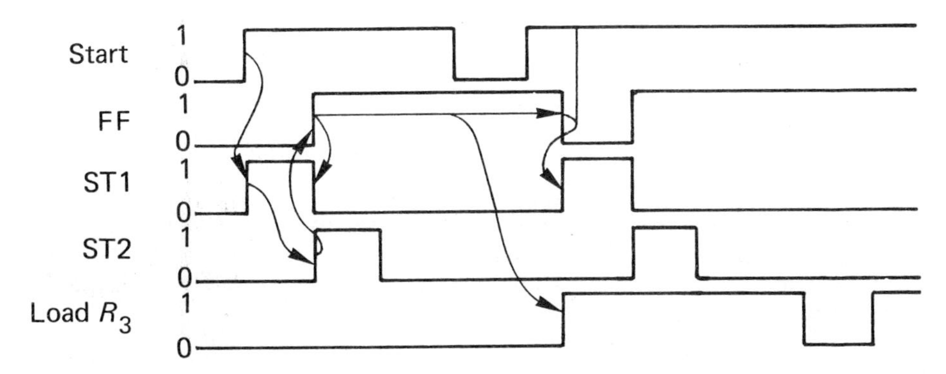

Fig. 1.2 Timing for Fig. 1.1

zero time, but in a time known as the **gate delay** (see below). The flip-flop, FF, cannot be reset and then set one gate delay later. There must be some minimum *time*, say WAIT2, between gate output and flip-flop input. Thirdly, for the process to be carried out as described here, the registers and FF must be affected by *edges*, and not by logic levels. If this is not so, then values at the input to registers and FF have to be held constant during a longer period of time, and things become yet more complex.

The following paragraphs will indicate some other aspects of system design that involve time.

Effect of capacitance

All wires and all electronic devices are constructed from conducting materials which have a finite area. Any voltage on a wire will therefore produce an electric field between the wire and earth, and between itself and other wires. This is represented in a circuit as a **capacitance**. Where the capacitance is not deliberately introduced, it is known as a **stray** capacitance.

Whenever an attempt is made to change the voltage on a capacitance, a certain amount of charge must flow into or out of it. The charge on a capacitance is given by the equation

$$q = Cv$$

where q, C and v are the charge, capacitance and voltage respectively. Thus

$$\frac{dq}{dt} = C\frac{dv}{dt} = i$$

since the rate of change of charge is a current, i. From this equation it can be deduced that the voltage change cannot be infinitely fast unless the current is infinitely large ($i \to \infty$ as $dt \to 0$). Thus the following rule can be stated.

It is impossible to change the voltage across a capacitance instantaneously

Since the capacitance cannot be charged in zero time by a finite current, all circuits, including connecting wire, introduce some time delay into a system.

Inductive effects

The flow of current will produce a magnetic field. By Lenz's Law, the field will tend to oppose the current flow. This effect is represented in a circuit by means of an inductance. For an inductance one can write

$$v = L\frac{di}{dt}$$

where L is the inductance, and hence

It is impossible to change the current in an inductance instantaneously

Inductive effects are usually small, unless either there is some ferromagnetic material present (e.g. magnetic core stores), or the wires are relatively long. What is meant by 'long' depends on the circuitry being used. It may be a few tens of millimetres where the fastest circuits are concerned, or it may be metres or even kilometres for slow circuits. A more precise definition will appear in Chapter 7.

Transmission line effects

Where the wires are relatively long they possess both capacitance and inductance. This leads to the theory of transmission lines as discussed in Chapter 7. The effect is that of the time for a signal to travel along the wire. The velocity of an electromagnetic wave (e.g. light) is $3 \times 10^8 \mathrm{m\,s^{-1}}$ in a vacuum, but usually only a half to two-thirds of that in a wire. This may seem very fast. Circuits operating at speeds well under 1 ns have been designed, and those operating at 3 ns are common (1983). As the velocity of a signal represents about 3 ns per metre, or 1 ns per foot, in vacuo, it will be seen that wires can easily make up an important part of the timing of a system.

Logic delays

The effects of capacitance are responsible for circuit delays. Within a particular logic function, all of these times are lumped together to give a single set of timing parameters for that function. Table 1.1 gives the times for a 74LS00 logic gate. This device is a two input NAND gate, of which there are four in a 14-pin package. The delay time is t_p, defined as the time between the input and corresponding output changes of the gate. The delay time for the rising and falling edges may be different, as is the case shown in Table 1.1. In these circumstances t_p is qualified by further subscripts indicating the high to low (HL) or low to high (LH) change of the output (Fig. 1.3).

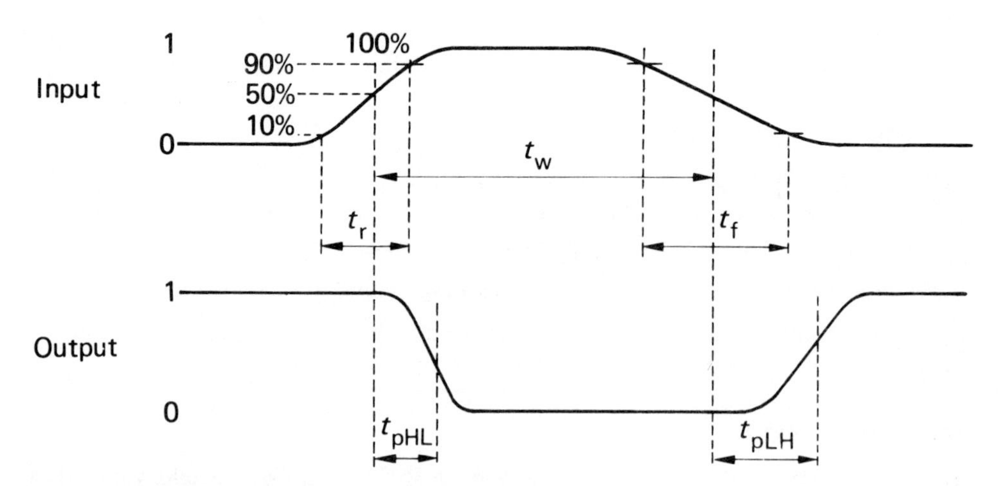

Fig. 1.3 Timing definitions

Table 1.1 Timing of the 74LS00 gate (all times in ns)

	Typical	Maximum
t_{pLH}	9	15
t_{pHL}	10	15

Logic families

There are several families of logic circuits on the market, usually distinguished by the technology of their manufacture. The most important of these are as follows.

Transistor–Transistor Logic (TTL)
A very popular family with a large range of functions. There are several sub-families with different speed and power consumption characteristics. Most of the devices are designated 74 . . . or 54. . . . In this book the 74LS . .* series will be used for examples. Switching times for a simple gate are shown in Table 1.1. The low state is about 0 V (often taken as 0.2 V herein), the high state (V) is between 3 V and 4.5 V (usually taken as 3.5 V), the switching threshold is about 1.3 V, and the power supply is 5 V. The output impedance in the low state is a few ohms, and in the high state around 120 Ω. The input impedance is a few kilohms, 22 KΩ in the LS series.

Emitter Coupled Logic (ECL)
A high speed family with a low output impedance in both logic states, and a high input impedance. Gate delays are generally less than 2 ns, and internally to the chips can be less than 0.5 ns. Output levels are −0.85 V and −1.7 V, with a switching threshold of −1.3 V. The power supply is usually −5.2 V.

Complementary Metal Oxide Silicon (CMOS)
This family is similar in function to TTL. Power consumption is almost nil except when actually switching between states, and the circuits will operate over a wide range of power supply voltages (e.g. 3 to 18 V). Circuits are generally slow, delay times being around 100 ns for an equivalent to the gate in Table 1.1, and edge times are of a similar order (5 V supply). High speed manufacturing processes are beginning to appear, giving delay times internal to a chip of less than 10 ns (1983). Output levels are the supply voltages, and the switching threshold is about half-way between.

Definitions relating to waveforms

Real waveforms do not have infinitely fast edges. Figure 1.3 shows a real pulse and its time parameters. The pulse width, t_w, is measured between the 'half amplitude' points. The pulse edge times cannot be measured accurately, as they are asymptotic to the final values. Hence edge times are normally specified between points at 10% and 90% of the amplitude, as shown in Fig. 1.3. The reader is warned, however, that 10% and 90% do not always apply. For ECL circuits in particular, rise and fall times are quoted between 20% and 80% of amplitude.

Figure 1.3. also shows the result of putting the pulse into a 74LS00 gate, and the definitions of t_{pLH} and t_{pHL}.

* LS implies Low power Schottky, a version of the technology operating at the same speed as earlier versions, but requiring less power.

Summary

This chapter indicates the importance of *time*, primarily in electronic systems. It points to some of the places where time may assume importance. The purpose of the remainder of this book is to show how timing can be calculated and controlled, in order to be able to construct reliable systems.

2 Basic concepts

Fundamental equations

It was stated in Chapter 1 that one of the causes of time delays in electronic circuitry is stray capacitance. Although this effect is in itself a nuisance, it can also be exploited. By artificially increasing the capacitance, the delay can be increased and used to shape and generate pulses, as will be described in later chapters. It is necessary, first of all, to study the basic circuits.

Figure 2.1 shows a circuit consisting of a pure voltage square wave generator, in series with a resistance, and driving a capacitive load. The resistance may be the internal resistance of the generator, or an external resistance, or a combination of both. The capacitance may be due to wiring, or to the inherent capacitance of some electronic device, or a deliberately introduced component.

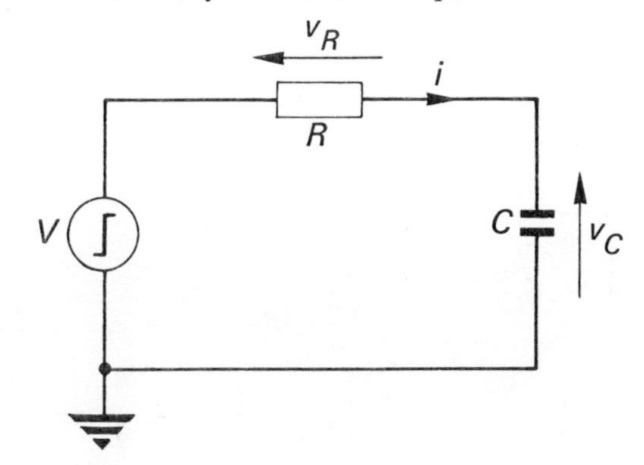

Fig. 2.1 Simple R–C circuit

Using Kirchoff's voltage law, the voltages in the circuit are added. Using the equation $i = C\mathrm{d}v_C/\mathrm{d}t$, this gives

$$V = v_R + v_C$$

$$= iR + \frac{1}{C}\int i\,\mathrm{d}t$$

Differentiating with respect to time

$$0 = R\frac{\mathrm{d}i}{\mathrm{d}t} + \frac{i}{C} \quad \text{(since } V \text{ is a constant)}$$

$$\frac{\mathrm{d}i}{i} = -\frac{\mathrm{d}t}{RC}$$

$$\ln i = -\frac{t}{RC} + \text{constant}$$

Suppose that the voltage is switched from 0 to V volts at time $t = 0$. The voltage across the capacitance is initially 0, as we cannot change the voltage across a capacitance instantaneously. Hence

$$i = V/R$$

and

$$\ln (V/R) = \text{constant}$$

$$\ln \frac{i}{V/R} = -\frac{t}{CR}$$

or
$$i = \frac{V}{R}\, e^{-t/CR}$$

and hence $\quad v_R = iR = V e^{-t/CR}$

and
$$v_C = \frac{1}{C}\int i\,dt$$
$$= -\frac{V}{CR} CR\, e^{-t/CR} \Big|_0^t$$
$$= V(1 - e^{-t/CR})$$

These two equations are of fundamental importance, and will occur repeatedly in one form or another. Figure 2.2 is a diagram of the three waveforms in the circuit. Two characteristics are important.

At time $t = 0$, $v_C = 0$ and $v_R = V$
At time $t = \infty$, $v_C = V$ and $v_R = 0$

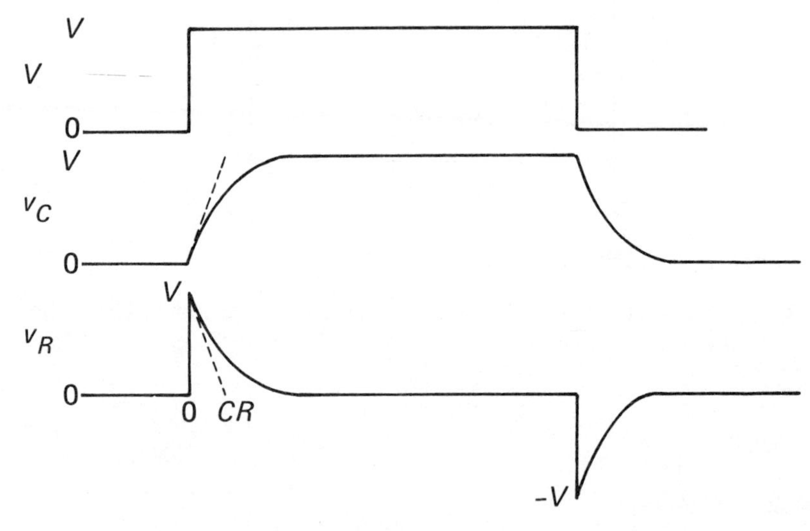

Fig. 2.2 Waveforms for Fig. 2.1

However, time never reaches infinity. Substituting values of $t = nRC$ into the equations (n integral), Table 2.1 is constructed. From this it is seen that the voltages v effectively reach their final values in a time between 3 and 5 times CR. The quantity

CR has the dimensions of time, and is known as the **time constant**. It can be shown very easily that if the initial slope of the voltage waveform were extended to cut the final value of the voltage, it would do so at time CR.

Table 2.1 Characteristics of an exponential waveform

t	*% of final value reached*
CR	63.2
2*CR*	86.5
3*CR*	95.0
5*CR*	99.3

If the rise time (or fall time) of a waveform is due to a time constant of this form, the rise time can be calculated as follows.

$$v = V(1 - e^{-t/CR})$$

$$V - v = Ve^{-t/CR}$$

$$e^{t/CR} = \frac{V}{V - v}$$

$$t = CR \ln \frac{V}{V - v}$$

Hence for $v = 0.1V \quad t = 0.105CR$
and for $v = 0.9V \quad t = 2.303CR$

Hence the rise time is $2.2CR$

One further fact is sometimes useful. If this circuit is analysed for its response to a sine wave, it will be found that, for a constant input magnitude, the output has a constant magnitude over a frequency range from DC upwards, until a 3 dB cut-off frequency is reached. This is the frequency at which v_C is $1/\sqrt{2}$ of its value at 'low' frequency, and is given by

$$f_c = \frac{1}{2\pi CR}$$

Since the rise time $t_r = 2.2CR$, it is possible to substitute and obtain the formula

$$f_c = \frac{0.35}{t_r}$$

This is useful in assessing various instruments. For example, an oscilloscope with a frequency response of 10 MHz will have an inherent rise time of

$$t_r = \frac{0.35}{10 \times 10^6}$$

$$= 35 \text{ ns}$$

This means that if an edge of effectively zero rise time (less than about 5 ns) is displayed on this oscilloscope, the rise time will be measured as 35 ns.

Integration

Consider the waveform v_C, shown in Fig. 2.2, in a circuit where the time constant is large compared to the time scale of interest. An example is where the frequency of a square wave is relatively high, as shown in Fig. 2.3. The voltage v_C now approximates to a straight line **ramp** (a waveform of the form $v = kt$, where k is a constant). The change of height of this ramp is proportional to the area under the square wave from the previous edge. Hence, neglecting constant terms, this waveform is an approximation to the integral of the input waveform with respect to time. The *R–C* circuit is thus often described as being an **integrating circuit**.

The details of the voltage values of the output are not as simple as they first appear. When the input is first changed from 0 to V, the output will rise to V'. The input then falls to 0 V, giving a change *from* V' of V''. Thus the change $V'' = V'(1 - e^{-t/CR})$. As V' is much smaller than V, the output will not return to 0V before the next input change. After a few cycles (greater than 5 time constants) the output will settle down to swing about an average of $V/2$, when the values of V and V' will be the same.

An equivalent to this will be found if the input waveform is a sine wave. A sine wave is characterised by a **frequency**, f, in Hertz (repetitions per second), and a **phase** relative to some reference. (The mains supply is a sine wave of 50 Hz.) If $2\pi f$ is very much greater than $1/CR$, then the phase of v_C is 90° behind that of V. The integral of a sine wave is the negative of a cosine wave, which, in turn, is a sine wave shifted by −90°.

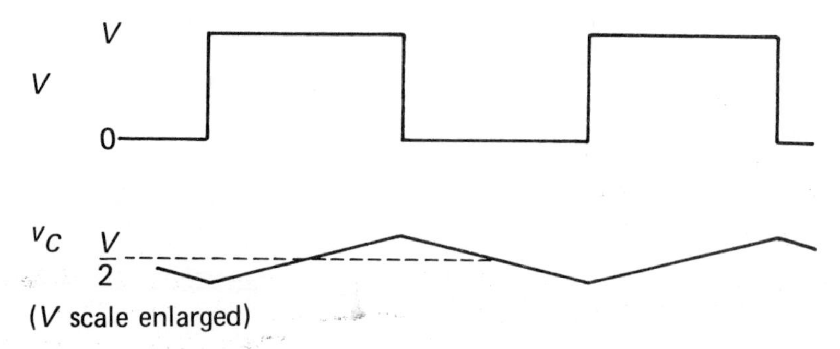

Fig. 2.3 Integrating action of the *R–C* circuit

Differentiation

Turning round the circuit of Fig. 2.1, the *C–R* circuit of Fig. 2.4 is formed. The values of v_R and v_C are identical to those of Fig. 2.1; only the 'earth' point has been moved. Figure 2.4 shows the voltage v_R.

If the time constant is small compared to the relevant time scale, then the waveform, v_R, appears as a set of spikes. The differential of an edge, or **step** such as appears as a change from 0 to V in the input waveform, is an **impulse**. An impulse has, ideally, an infinite height, zero width, and finite area (the integral is a step). The voltage v_R is thus an approximation to the differential of the input with respect to time. The circuit is known as a **differentiating circuit**. In the same way as for the integrator, an input sine wave whose frequency is *low* compared to $1/CR$ will result in v_R being +90° out of phase with the input, and is thus the differential of the input.

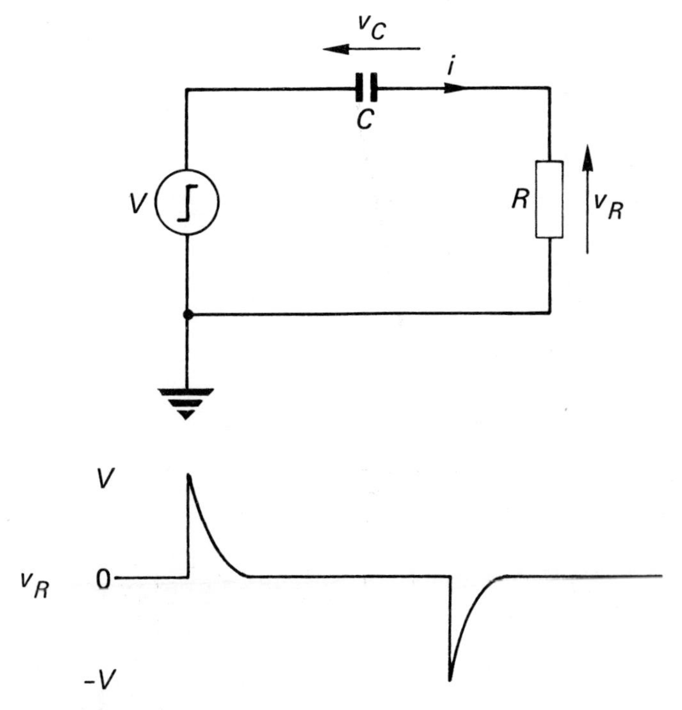

Fig. 2.4 The *C–R* differentiating circuit

Definition of time periods

The exponential waveforms produced by *C–R* and *R–C* networks can be used to define time intervals. Clearly, the complete exponential cannot be used, since the rate of change of voltage close to the end of the rise is very small. It is thus difficult to define the time accurately with real components. In practice, cheap, easily available components often have tolerances of 5 or 10%. Figure 2.5 illustrates this. In Fig. 2.5(a) the rate of change of v is high at the end of the required time period. If the tolerance on the end of the time period is ±10 mV (say), then the error in the time is fairly small. Close to the end of the exponential, however (Fig. 2.5(b)), the same tolerance on voltage levels gives a large error in the time.

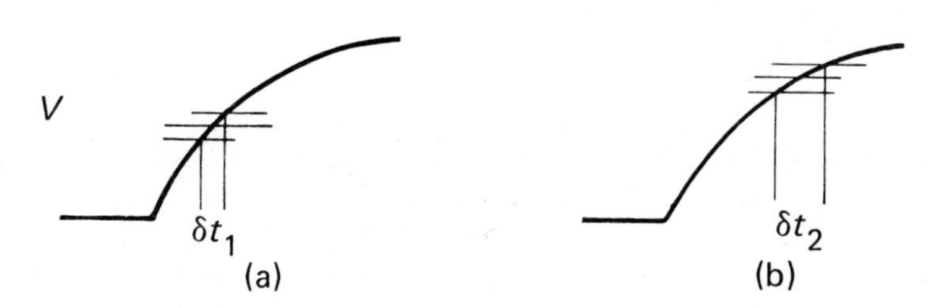

Fig. 2.5 Accuracy of time definition

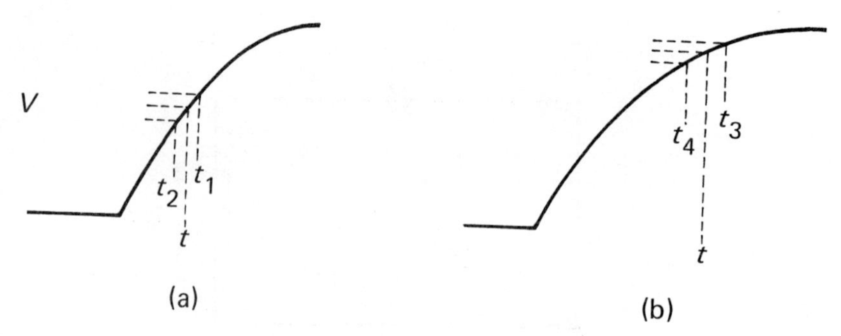

Fig. 2.6 End of a time period

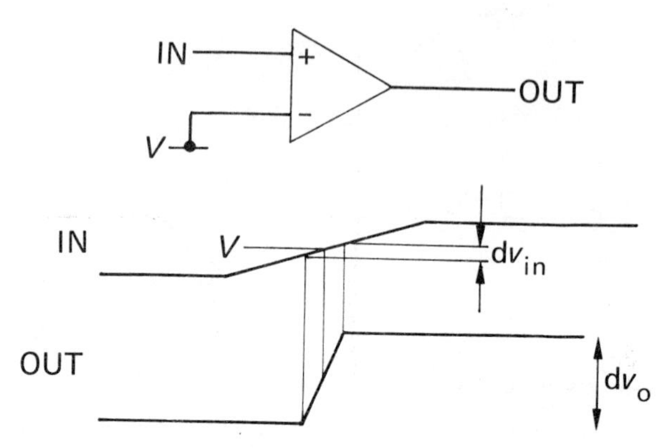

Fig. 2.7 Comparator and its operation

Put mathematically, the time t for the voltage to reach v is given by

$$t = T \ln \frac{V}{V - v}$$

where T is the time constant. Suppose $v = V/2 \pm V/20$ (i.e. a tolerance of 5% of full swing), as shown in Fig. 2.6. Then

$$\delta t = t_1 - t_2 = T \ln \frac{11V}{9V}$$

$$= 0.20T$$

If, instead, v was $0.9V$ with the same tolerance, then

$$\delta t = t_3 - t_4 = T \ln \frac{3V}{1V}$$

$$= 1.10T$$

To define the end of the time period it is necessary, for a very small change at the appropriate point on the exponential, to cause a digital voltage change, as shown in Figs 2.6 and 2.7. To achieve this, the exponential waveform is used as input to a high gain amplifier. An amplifier is a device that accepts a small change at the input, and produces a larger change at the output. The ratio dv_o/dv_{in} is known as the **gain**. Because of the large value of the gain, a large change of input will take the amplifier

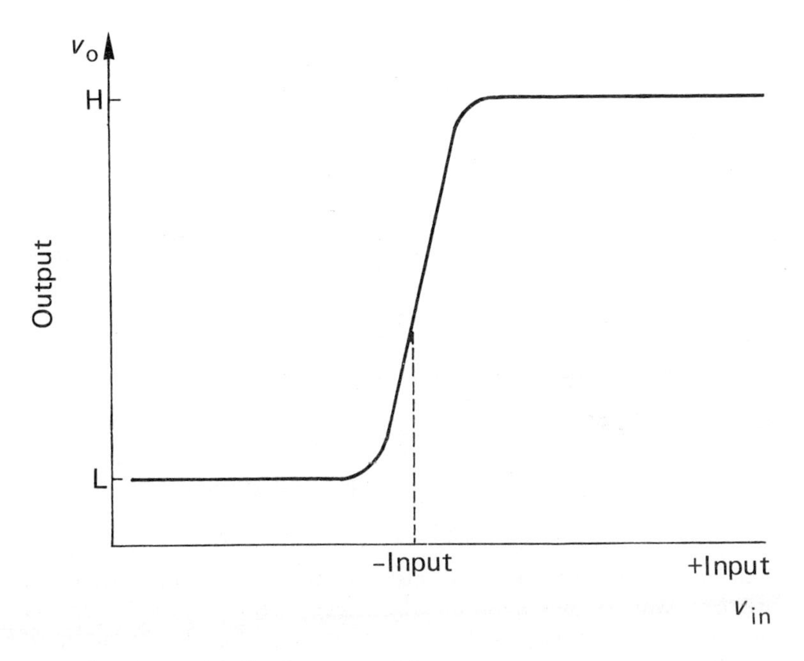

Fig. 2.8 Transfer characteristic of a comparator

output to one of the limits of its operation. The amplifier here is such that if the '+' input is at a higher voltage than the '−' input, then the output goes hard to the higher limit of the output voltage range. If the '+' input is below the '−' input, then the output goes to the lower limit. The output is, therefore, digital in nature. Only when the two inputs are approximately equal is the output *between* the digital states, and it is in this area that the gain is high. The type of amplifier is called a **comparator**. The exponential waveform is made to start from a relatively low voltage, and aims for a relatively high voltage (or vice versa). The '−' input of the comparator is set about half way between. When the waveform passes this voltage, the comparator output changes state.

It should be realised that most digital circuits operate for short periods in a linear manner, i.e. the output is proportional to the input. The period concerned is during switching from one of the two digital states to the other. For example, a comparator input may need to change by 10 mV in order for the output to change fully between its two states. Figure 2.8 is a graph of input against output. The gain of the device, expressed as dv_o/dv_{in}, is low or high. In switching between the two output states the gain is high, and there is a linear relation between input and output, i.e.

$$v_o = Av_{in}$$

For a simple silicon transistor, a change of voltage of about 120 mV between the base and the emitter will cause the collector current to change by a factor of 100 in the switching region. (This is a useful figure to bear in mind.) When not switching between the two digital states, the gain of the comparator of Fig. 2.8 is low, the output being virtually constant for input voltage changes.

More complex *R–C* circuits

The simple circuits of Figs 2.1 and 2.4 are rarely a true representation of reality, but are quite often a good approximation to it. Where this is not the case, two other

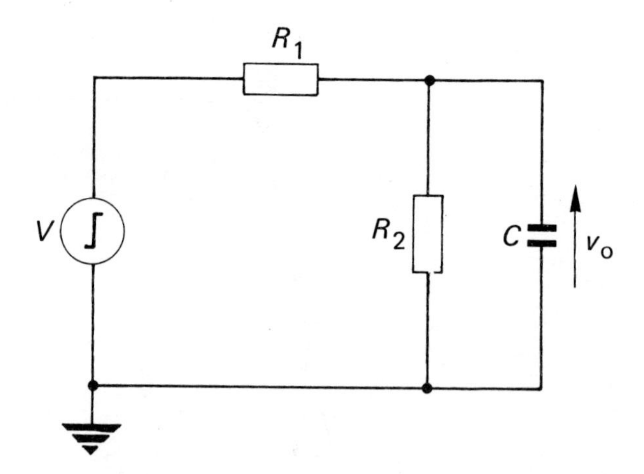

Fig. 2.9 Resistance in parallel with capacitance in the *R–C* circuit

circuits cover the vast majority of situations. The first of these is shown in Fig. 2.9.

There are three ways of understanding the effects of this circuit. The first is to perform an integration comparable to that performed earlier for the circuit of Fig. 2.1. This is rather tedious to perform, and will not be attempted here. It gives the same results as the methods described below, but readers not familiar with linear electronics may like to confirm that.

The simplest, but possibly the hardest method to accept, is to realise that a voltage generator is an 'AC short circuit'; that is, it is effectively zero impedance to a signal of the frequency of interest. Fourier analysis shows that a square wave contains frequency components up to very high frequencies; infinity for infinitely fast edges. If this short circuit is drawn in, then it is clear that the R on Fig. 2.1 becomes R_1 and R_2 in parallel in Fig. 2.9. Thus v_o in Fig. 2.9 will change on a time constant $CR_1R_2/(R_1+R_2)$. The initial value of v_o will still be zero, since the voltage across a capacitance cannot be changed instantaneously. The initial current is therefore V/R_1. The final value of v_o is determined by the two resistors, since there is no current in the capacitance. Hence

$$v_o = \frac{VR_2}{R_1+R_2} \quad \text{and} \quad i = \frac{V}{R_1+R_2}$$

The third method of determining this effect is to perform an AC analysis using the jω technique. This leads to the expression

$$v = \frac{VR_2}{R_1+R_2} \frac{1}{1+\dfrac{j\omega CR_1R_2}{R_1+R_2}}$$

The situation of Fig. 2.1 is represented by setting $R_2 = \infty$, giving

$$v_o = \frac{V}{1+j\omega CR_1}$$

The time constant, CR_1, appears in this formula. By analogy, the circuit of Fig. 2.9 will have a time constant $CR_1R_2/(R_1+R_2)$, R_1 having been replaced by R_1 in parallel with R_2. The first term in the expression gives the final voltage.

The second of the two circuits of importance is illustrated in Fig. 2.10. It is found most frequently as the 10× probe of an oscilloscope, where $R_1 = 9R_2$, and C_2 is the input capacitance of the oscilloscope amplifier. C_1 is an adjustable capacitance. The

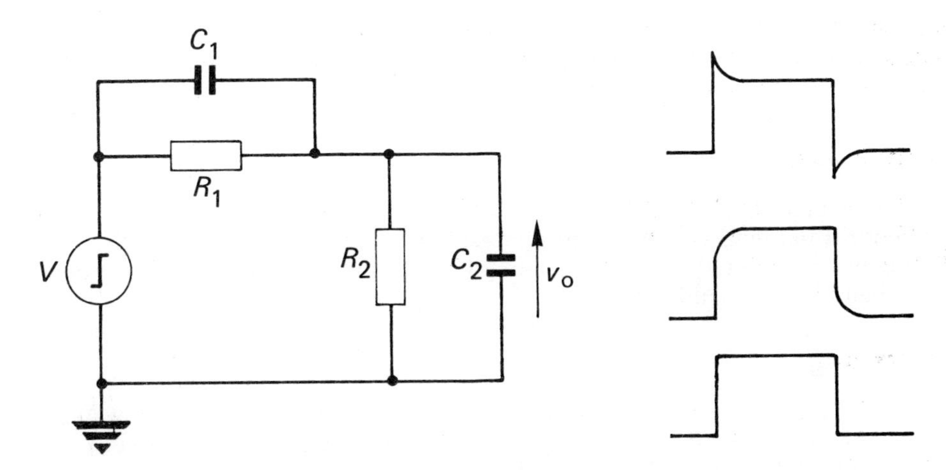

Fig. 2.10 Circuit of a 10X probe of an oscilloscope

effects are not simple. There are two capacitances in series and, since the voltage across a capacitance cannot change instantaneously, the input edge cannot be infinitely fast. It can, however, be fast relative to the time constants. It would therefore be reasonable to suppose that the initial value of v_o would be determined by the capacitance alone. Hence

$$v_o = \frac{\dfrac{1}{j\omega C_2}}{\dfrac{1}{j\omega C_1} + \dfrac{1}{j\omega C_2}} \times V = \frac{VC_1}{C_1 + C_2}$$

The final value will be determined by the resistance alone. Hence

$$v_o = \frac{VR_2}{R_1 + R_2}$$

If these two values are equal, then the output will be an almost faithful replica of the input. Thus

$$\frac{C_1}{C_1 + C_2} = \frac{R_2}{R_1 + R_2}$$

$$C_1(R_1 + R_2) = R_2(C_1 + C_2)$$

$$C_1R_1 = C_2R_2$$

This is a slightly surprising result.

Figure 2.10 shows the waveform of v_o in three cases: where the initial value of output is greater than, less than or equal to the final value. The time constant can be calculated using the second or third of the methods used for the circuit of Fig. 2.9, and is $(C_1 + C_2)R_1R_2/(R_1 + R_2)$.

Circuit delays

All circuits introduce time delays between input and output. The causes of this can be explained in different ways, but all reduce to the effects of capacitances. The details of the internal workings of integrated circuits are beyond the scope of this text. It is necessary to be aware of these delays, however, and they will be illustrated, usually exaggerated, at relevant points. Even where they are not shown, the reader should realise that they do exist.

Conclusion

This chapter covers the basic topic of *R–C* and *C–R* circuits. Using this knowledge, the majority of timing circuits can be understood. Once this chapter has been thoroughly mastered, the rest of the book should be little more than examples of its use, albeit sometimes moderately complex.

Questions

1 The input circuit of a system consists of a series resistance of 200 Ω, followed by a capacitor of 50 pF (see Fig. 2.1). If the driving circuit is a pure voltage source, what is the rise time at the output of this circuit?

2 A square wave of frequency 10 MHz and of 1 V amplitude is fed to the circuits in Fig. 2.11. Sketch the output waveforms in both cases, assuming that the square wave has been running 'for a long time'. Repeat for frequencies of 100 KHz and 1 KHz.

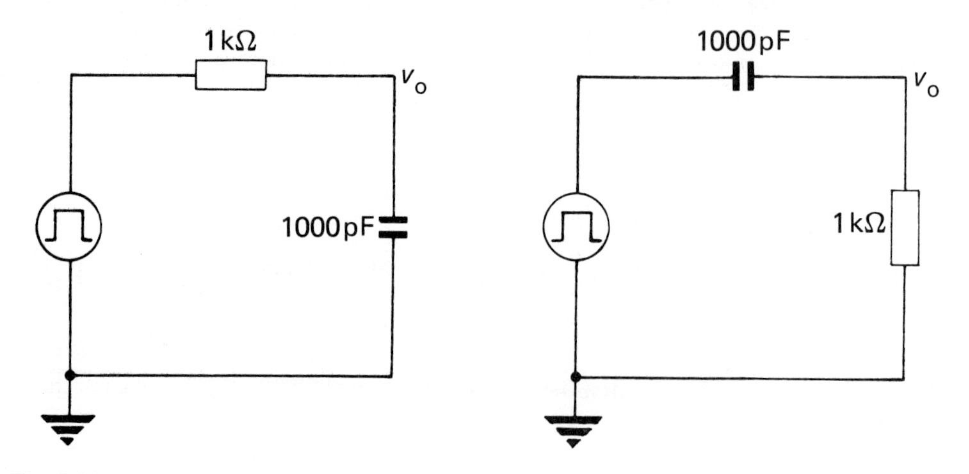

Fig. 2.11

3 A TTL gate drives a load that is equivalent to 50 pF and a high resistance in parallel. The output impedance of the gate is 100 Ω on the low to high edge. What will be the rise time of the signal?

4 Assuming diode resistance to be incorporated in the 20 Ω source resistance, calculate the rise and fall times of v_o in the circuit of Fig. 2.12.

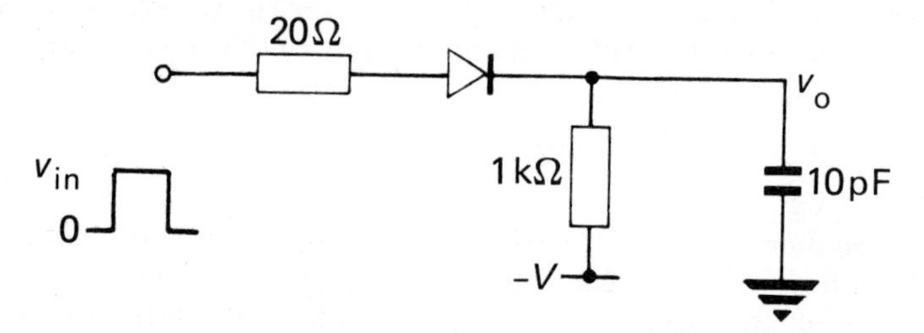

Fig. 2.12

5 An oscilloscope input impedance consists of 1 MΩ in parallel with a 72 pF capacitor. A 10× probe is used, causing a series resistance of 9 MΩ to be added to the circuit (see Fig. 2.9). If an infinitely fast edge is applied to the probe, and the amplifiers of the oscilloscope are very fast, what rise time would be observed?

6 A capacitor is added in parallel with the 9 MΩ resistance. What value is needed to obtain a faithful reproduction of the waveform at the oscilloscope input?

7 If this capacitor was 6 pF, sketch the waveform at the oscilloscope input.

8 Repeat Question 7 for a capacitor of 10 pF.

3 The Schmitt trigger and feedback

The need for a hysteresis characteristic

In most timing circuits the time period is determined by a delay generated by an R–C or C–R network. To achieve accurate timing, the amplifier switching point must be on the steep part of the exponential change. Even so, where long time periods are concerned, the amplifier input will spend a significant time in the linear region. Any departure from the exponential will appear in amplified form at the output. As the amplifier will generally have a high input impedance, it is easy to pick up extraneous signals (noise) which will cause such disturbances. In a bad case the output may oscillate.

Consider a device having a transfer characteristic as shown in Fig. 3.1. When the input voltage, IN, is low*, the output is at V_H, the higher of the two logic levels. As the input voltage rises, the output stays sensibly constant until a switching *threshold* is reached. The output now falls to V_L, the lower of the logic levels. If the input is *reduced* slightly below this switching threshold, the output voltage does not change. Indeed, the input needs to fall some way before the output will switch back to V_H. This characteristic looks rather like the hysteresis loop of a magnetic material, and hence its name. It may be compared to backlash in a mechanical system.

Returning now to the problem of noise on a slowly changing input to this device, it

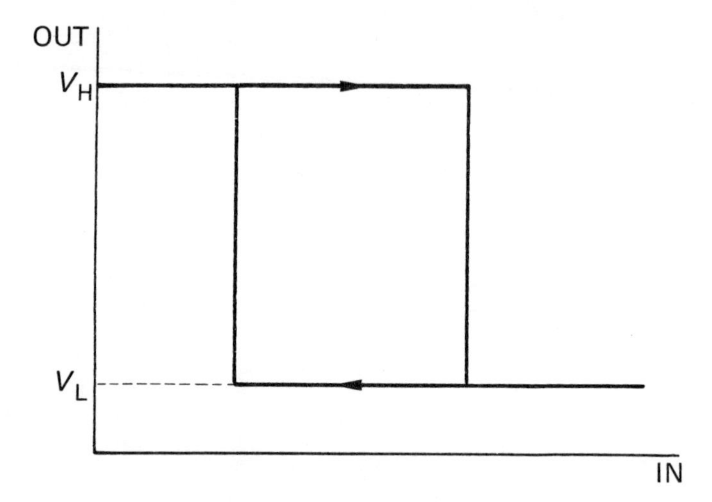

Fig. 3.1 Transfer characteristic with hysteresis

*The terms 'high' and 'low', 'positive' and 'negative', are used in this text in a relative sense, and not absolute. Thus, for a TTL gate the voltage levels 0 and 3.5 V may be referred to as low and high, or negative and positive, respectively. Positive going and negative going are defined analogously.

will be seen that once the output has switched, the noise will have no effect, provided that its amplitude is held within quite *un*restrictive limits. The 'vertical' parts of the transfer characteristic still need to be steep, of course, implying high gain. This can be improved by the use of positive feedback.

Feedback

The theory of feedback systems is extensive, but the information given below is sufficient for this text.

Feedback means, quite simply, taking the output of a circuit and feeding part or all of it back to the input. If the signal fed back is in the same sense as the input, the feedback is positive. If it is in the opposite direction, the feedback is negative.

Consider the circuit of Fig. 3.2. A fraction, β, of the output is added to the input. Suppose v_o is low. IN goes sufficiently high to cause v_o to begin switching. The feedback adds to IN, increasing the input to the comparator, causing a larger output, causing a larger input, ... The output, therefore, snaps over very rapidly.

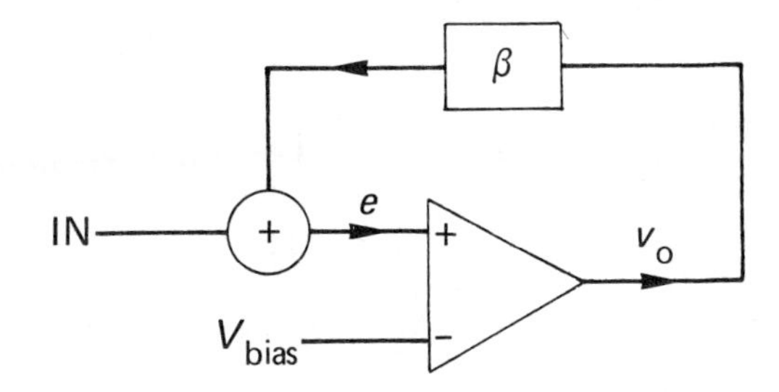

Fig. 3.2 Positive feedback round a comparator

Within a region where the gain of the comparator, $A = v_o/e$, is constant,

$$v_o = Ae$$

$$e = \beta v_o + v_{in}$$

$$v_o(1 - A\beta) = Av_{in}$$

$$v_o = \frac{Av_{in}}{1 - A\beta}$$

If IN is low, then v_o is at the negative limit. As IN tends to V_{bias} the gain, A, rises and $A\beta$ tends to unity. Hence v_o/v_{in} tends to infinity. The gain will never become negative, since the circuit would reach another part of its characteristic where A is lower before v_o reached 'infinity'.

If the comparator output is fed back to the '−' input, then

$$v_o = \frac{Av_{in}}{1 + A\beta}$$

and if $A\beta$ is very large, $v_o \simeq v_{in}/\beta$. As β is usually made from passive components, these define the circuit gain independently of the value of A.

The Schmitt trigger

Figure 3.3 shows one form of Schmitt trigger circuit. If v_{in} is lower than the '+' input of the comparator, v_A, then v_o is at V_H, the higher logic level. Hence

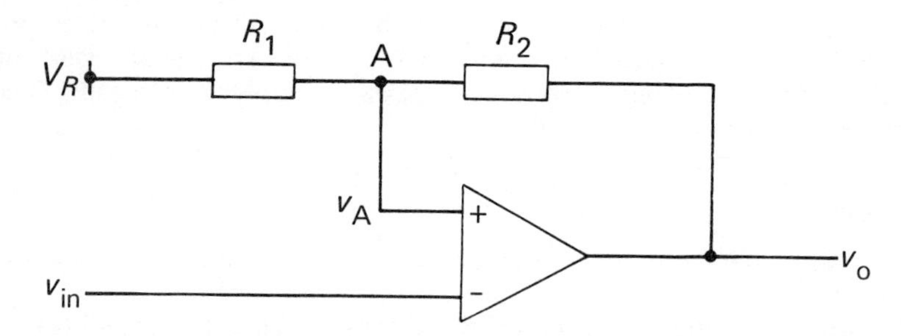

Fig. 3.3 Inverting Schmitt trigger circuit

$$v_A = V_R + \frac{R_1}{R_1 + R_2}(V_H - V_R)$$

and the high threshold, V_{THH} is given by

$$V_{THH} = \frac{R_2 V_R + R_1 V_H}{R_1 + R_2}$$

As the input rises v_o, and hence the threshold voltage V_{THH}, remains constant until v_{in} reaches V_{THH}. As the rise continues, v_o falls by an amplified version of the small change of v_{in}, and this appears as a fall in v_A, making the change in comparator input voltage large. This is the positive feedback. The output eventually reaches the lower logic level, V_L, and there is a new threshold voltage, V_{THL}, given by

$$v_A = V_{THL} = \frac{R_2 V_R + R_1 V_L}{R_1 + R_2}$$

As V_L is less than V_H, V_{THL} is less than V_{THH}. The transfer characteristic is thus like that in Fig. 3.1.

Figure 3.4 shows an alternative circuit. In this case a low input corresponds to an output at V_L. The characteristic will therefore be different from that of Fig. 3.1. With input low, v_A will be given by

$$v_A = v_{in} + \frac{R_3}{R_3 + R_4}(V_L - v_{in})$$

$$= \frac{R_4 v_{in} + R_3 V_L}{R_3 + R_4}$$

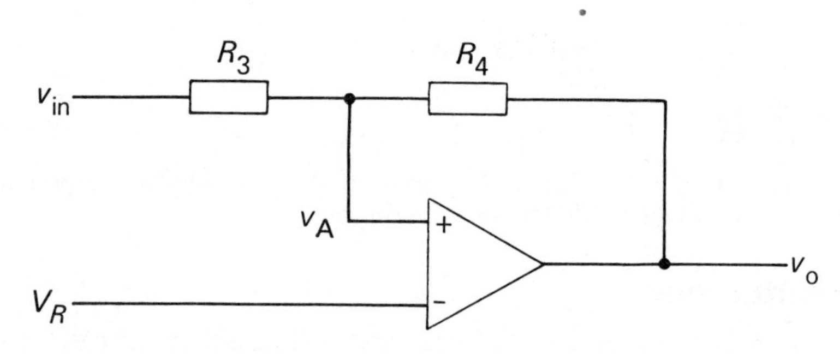

Fig. 3.4 Non-inverting circuit

When v_A becomes just greater than V_R, v_o will rise by an amplified version of the difference, forcing v_A higher, and causing fast switching by positive feedback. When the output reaches V_H, v_A will have changed to

$$v_A = \frac{R_4 v_{in} + R_3 V_H}{R_3 + R_4}$$

Now v_{in} will have changed very little, and as V_H is larger than V_L, v_A will have risen substantially. Thus it will be necessary for v_{in} to fall some considerable way before v_A falls to V_R, and the output is able to switch back. It is left as an exercise to the reader to draw the transfer characteristic.

The 74LS13

The TTL circuit 74LS13 is a Schmitt trigger circuit. Figure 3.5 shows the essential part of it. Other versions of the 74×13 and 74×14 are similar.

The heart of the circuit is the loop formed by T1 and T2. The switching thresholds are 1.6 and 0.8 V. Let the input of the circuit be high. The input diode will be off, so the base of T1 rises, turning the transistor on. The collector will therefore be 'low'. The circuit is designed for T1 to saturate; that is to say, the collector and emitter voltages are almost the same. In practice, the collector voltage will be about 0.2 V higher than the emitter for this type of transistor. A significant amount of time is required to remove the saturation effects when switching off. To avoid this, a Schottky diode is connected between the collector and the base. Such diodes have an especially low voltage drop when conducting current, say 0.4 V instead of the more common 0.7 V. The collector–emitter voltage is now limited to 0.3 V, thus preventing saturation. By choosing an appropriate ratio of r to R, the common emitter is set at 0.8 V. As there is only about 0.3 V between the collector and emitter of T1, the base emitter voltage of T2 will be 0.3 V only, holding it off. T3 will be on.

This state will remain until the input falls sufficiently to start to turn T1 off at 0.8 V. As T1 turns off, its collector rises, turning T2 on. The common emitter is pulled up in turn, and hence turns T1 off faster. This is the positive feedback. The collector of T2

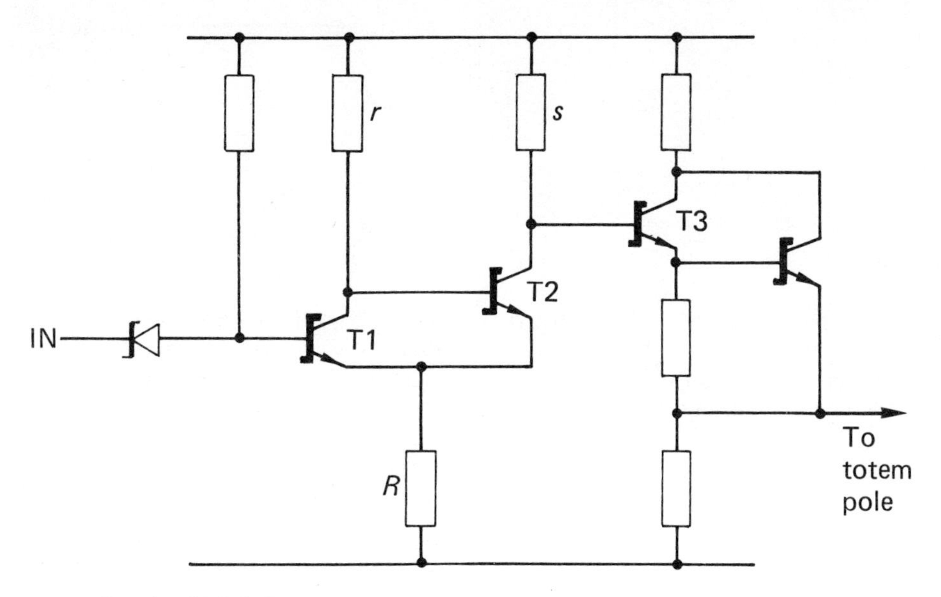

Fig. 3.5 Circuit of 74LS13

will fall. The final condition will be determined by the ratio of R to s. The common emitter is arranged to be at 1.6 V, and the T2 collector at 1.9 V due to almost saturating. At this point T3 is off. The state of the circuit will remain constant until the input rises sufficiently to start turning T1 on again, which will be when it reaches 1.6 V.

Summary

This chapter outlines an idea and a circuit, both of great importance. The idea of feedback is fundamental to a great deal of electronics, and the basics described will be used repeatedly in subsequent chapters. For the reasons expressed at the start of the chapter, the Schmitt trigger will also keep appearing.

Questions

1 Design a Schmitt trigger circuit with input switching thresholds differing by 1 V. The circuit is to work in association with TTL circuitry, and should have an output compatible with TTL levels. An appropriate comparator may be used if thought suitable.

Determine the actual threshold levels with the components chosen.

In what circumstances is it essential to use a Schmitt trigger circuit, rather than a conventional gate? U of M (June 1983)

[Assume TTL levels of 0 and 4 V; actual thresholds are calculated assuming real components, which will not allow exactly 1 V difference.]

Do this question using the circuits of Figs 3.3 and 3.4.

2

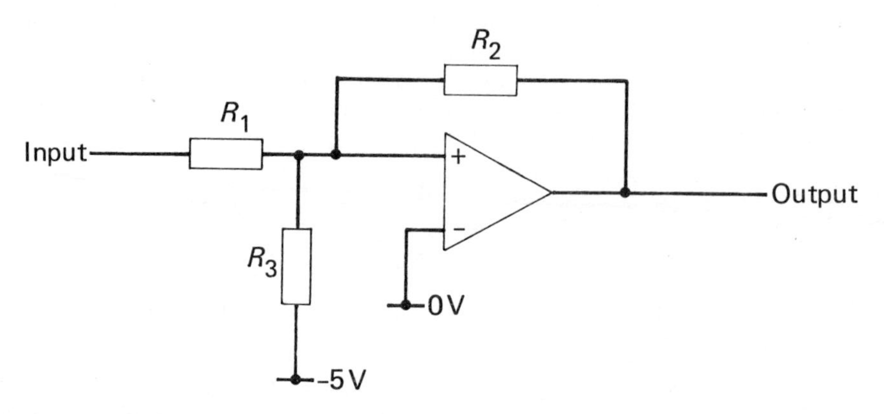

Fig. 3.6

The comparator shown in Fig. 3.6 gives output logic levels of 3.5 V and 0.2 V. The output is required to go high for input greater than +4 V and low for input less than +1 V. Choose values for R_1, R_2 and R_3 to achieve this with a maximum input current of 1 mA when the input is at +4 V.

4 Pulse shaping circuits

Introduction

There are several types of pulse shaping and pulse generator circuits available, and at least three types are needed. In particular, it is possible to distinguish between those which can only produce pulses shorter than the input pulse (pulse shaping), and those which can, when required, produce an output longer than the input (pulse generators). The circuits to be described in this chapter all produce pulses shorter than the input trigger, and are unable to produce longer pulses. Some of these circuits make use of R–C networks; others use transmission lines. In one sense, the R–C integrator circuit can be thought of as a simulated delay line built from discrete components.

A basic circuit

Figure 4.1 shows the simplest circuit. When transients have died away, the transistor is on, since current flows to the base via R. The circuit should be designed to saturate the

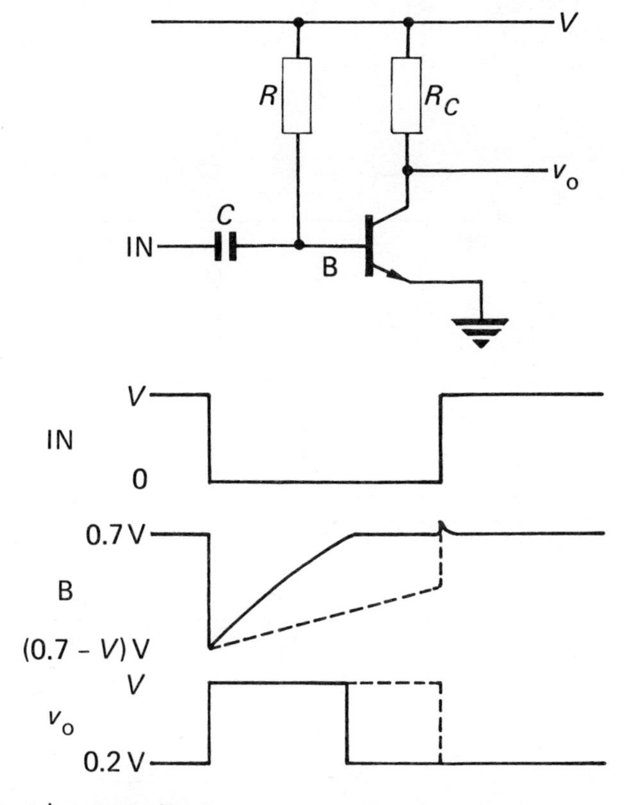

Fig. 4.1 Simple pulse generator

transistor, thereby avoiding variations due to wide tolerances on β. Alternatively the same design process is used and Schottky diodes are added to the circuit to prevent saturation. This would be the process used in the 74LS and similar circuit families.

A positive going edge at the input will cause the transistor to turn on harder, but does not affect the output. A negative going edge causes the transistor to turn off. The base then charges up on a time constant CR. Since the transistor must saturate,

$$\frac{(V-0.7)\beta}{R} > I$$

where I is the current to the collector from both R_C and any load. The base-emitter voltage drop when the transistor is turned on is 0.7 V.

Assuming that the negative going input swing is also approximately V, the equation of the exponential is

$$V = (V - 0.7 + V)(1 - e^{-t/CR})$$

$$t = CR \ln \frac{2V - 0.7}{V - 0.7}$$

$$\simeq CR \ln 2 \qquad \text{if } V >> 0.7\,\text{V}$$

$$= 0.69CR$$

The dotted portion of the waveform in Fig. 4.1 shows what happens if the time constant is too long: the output is then the same length as the input. It should be clear from this why the output pulse cannot be longer than the input.

Use of a transmission line

An alternative method of generating a pulse is to use a transmission line as a delay circuit element. The delay element is shown as a 'sausage' shape in Fig. 4.2. The small delays between A and B, between B and the falling edge of D, and between C and the

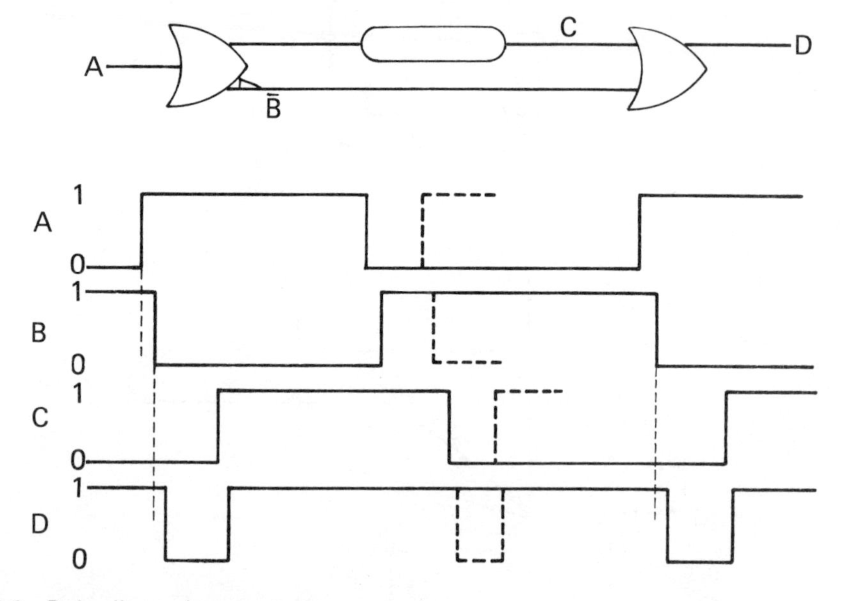

Fig. 4.2 Delay line pulse generator

rising edge of D, represent the inherent delays of the OR gates. The width of the pulse is equal to the length of the delay element. However, few gates have both true and inverse outputs. Only ECL circuits have this facility, particularly when changes at almost the same time are concerned. Where an extra gate must be inserted, the delay due to one gate will have to be added to or subtracted from the pulse width, depending on where the gate is inserted.

There are limitations on this circuit. Firstly the gate driving the delay must be capable of doing so. ECL gates can be matched to a discrete transmission line, but TTL cannot (see Chapter 7 for reasons). However, some specially buffered TTL compatible lines in DIL packages have recently become available. Secondly, whilst these special TTL lines may have delays of hundreds of nanoseconds, discrete delay lines are very bulky. One metre or so would be the practical maximum, giving a delay of about 10 ns. A line printed on a miniature substrate can have very small dimensions, and is therefore much more compact, but its quality is generally poorer—the rise and fall times of the output edges will be much worse than the input edges, and the amplitude may deteriorate due to the series resistance of the thin wire. Thus pulse lengths of the order of 50 ns are the longest that can be generated.

Recovery time

This circuit also suffers from a recovery time problem. In Fig. 4.2 the dotted waveforms show what happens when the low period of the input is shorter than the delay. B also has only a short pulse, and the result is that the output pulse is determined by the length of the low period of the input, rather than by the delay. If the pulse is to be of the required length, the circuit must be allowed to **recover**. The time required is the **recovery time**. For this circuit the recovery time is equal to the pulse width.

Use of *R–C* integrator—TTL

Another way of forming a delay is to use an integrating circuit, as was mentioned in Chapter 2. Since it is not possible to change the voltage across a capacitor instantaneously, a little time must elapse before the output waveform reaches the switching threshold of the driven gate. Thus, there is a delay.

Figure 4.3 shows the circuit and waveforms. The gates may be TTL or ECL, but for the present purpose TTL gates will be assumed. Because of the *R–C* network, point B rises and falls on a time constant. This delays the switching of the second gate, so waveform D is delayed from A. In general, the delay on the two edges will be different. The second gate is an inverting gate, and its output is combined with the system input.

Two points are noted. The waveform, A, is used as input to the final gate, rather than using the output of the first gate. TTL circuits have a relatively high output impedance of 120 Ω or so in the high state. The output of the first gate will thus have a slow rising edge, and should not be used to drive the final gate.

The second gate is a Schmitt trigger. This is because any noise on the waveform at B, possibly due to an imperfect waveform at A, would cause multiple output changes. The hysteresis action prevents this.

Suppose the output of the first gate is low. The power supply to a TTL circuit is 5 V, and the input resistance of the Schmitt trigger is 20 kΩ (see Fig. 4.4). Assuming a low output of 0.2 V, and a V_{be} of 0.7 V, there is effectively 4.1 V across the two resistors. Thus the voltage across R is given by

$$v = \frac{4.1R}{R+20}$$

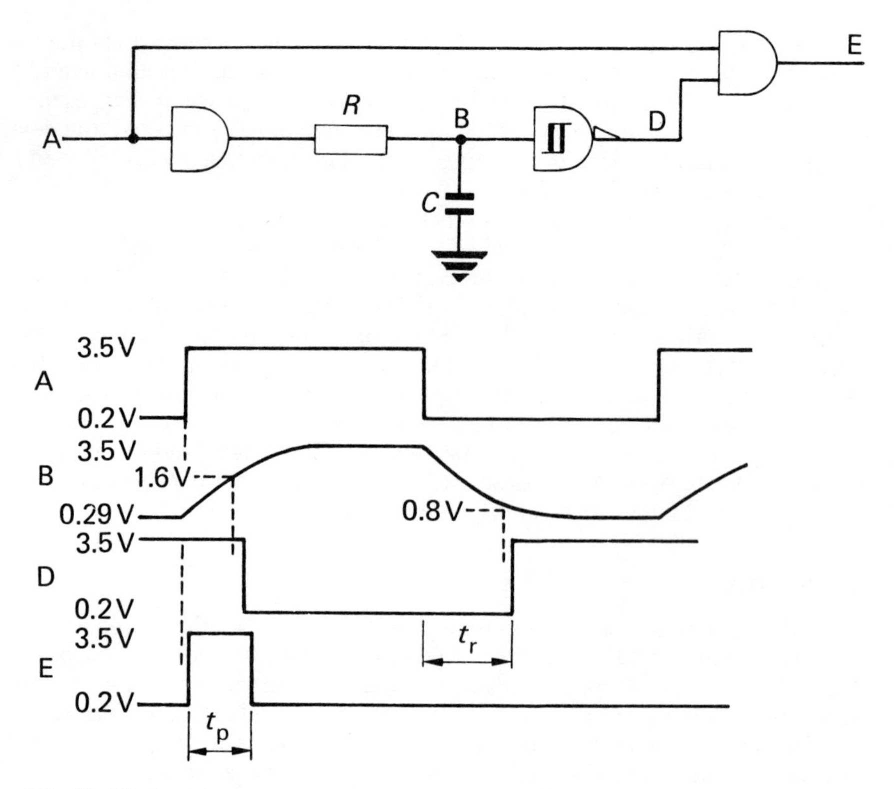

Fig. 4.3 *R–C* integrator

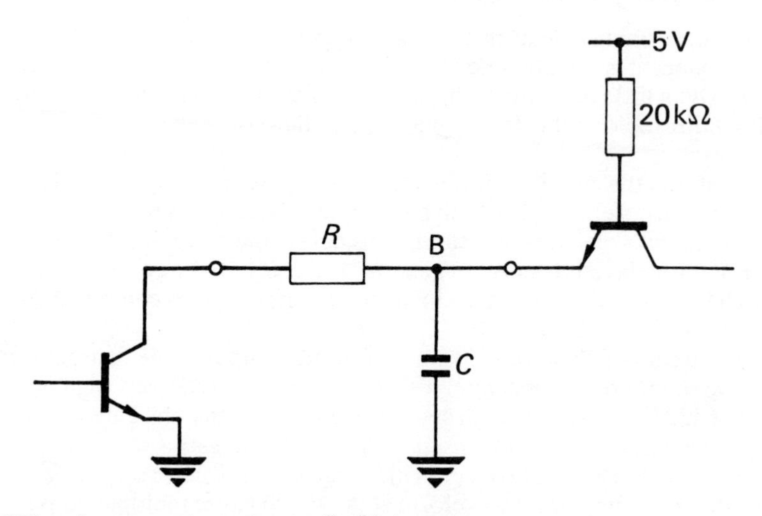

Fig. 4.4 TTL pulse generator using an *R–C* integrator

The switching points of the Schmitt trigger are at 0.8 V and 1.6 V. The static voltage at B must be below 0.8 V, or switching to the low state of D would never occur. It should be well below, or the delay on the falling edge of B will be very long, leading to a long recovery time. Suppose B to be required below 0.5 V. Then

$$\frac{4.1R}{R+20} < 0.5 - 0.2$$

$$3.8R < 6.0$$

$$R < 1.6\text{k}$$

Suppose R is 500 Ω. If the output voltage in the low state is 0.2 V, the low voltage at B will be 0.3 V. The higher switching threshold of the Schmitt trigger circuit is 1.6 V, and the high voltage output of most TTL circuits is measured at about 3.5 V typically. Assuming the output impedance of the driving gate in the high state to be 120 Ω,

$$1.3 = 3.2(1 - e^{-t_p/620C})$$

$$t_p = 620\,C \ln\frac{3.2}{1.9}$$

$$= 325\,C$$

This gives an output pulse of t_p.

By comparison with the previous circuit, the recovery time is equal to the delay. But the delay on the other edge is different. If the switching threshold of the Schmitt trigger is 0.8 V, then

$$(3.5-0.8) = 3.2(1 - e^{-t_r/500C})$$

$$t = 500\,C \ln\frac{3.2}{0.5}$$

$$= 920\,C$$

since the output impedance of the gate in the low state is negligible (say 10 Ω). Notice that this is almost three times the pulse width! The reason is the proximity of the switching threshold to the 'final' value (0.8 V to 0.3 V).

This circuit suffers from one very severe problem. The figures quoted in the above calculation are all 'typical'. The gate low output is specified as between 0 and 0.4 V, and the high output may be anything from 3 V to 4.5 V. The switching thresholds also have a tolerance. The value of the low threshold, in particular, is from 0.5 V to 1.0 V, and the low voltage at the input is 0.1 V above the low output of 0 to 0.4 V. Output and input impedance may vary by 20%. Thus, the times calculated are very approximate.

In spite of this objection, the circuit is widely used. The capacitor value is either selected for each circuit, or chosen to be such that the system will still work, even with the worst case tolerances. The circuit without the final gate is also used as a delay circuit on its own. It may be used where one or other edge delay is important. Or it may be designed so that the two edges are fairly well matched. Several iterations of the design will be needed to achieve this.

Use of *R–C* integrator—ECL

Figure 4.5 shows a pulse generator using ECL gates. The power supply to ECL circuits is −5.2 V, the logic levels are −0.85 V and −1.7 V, and the switching threshold is −1.3 V.

The output of an ECL circuit is driven by an emitter follower. An emitter follower has the characteristic that the base emitter junction behaves rather like a diode. When the base rises, the emitter *follows*. When the base falls, the emitter will follow if the load is resistive. If, however, there is a capacitive load, and the time constant is long enough, the diode will turn off. Hence the signal at B in Fig. 4.5 will fall slowly. The

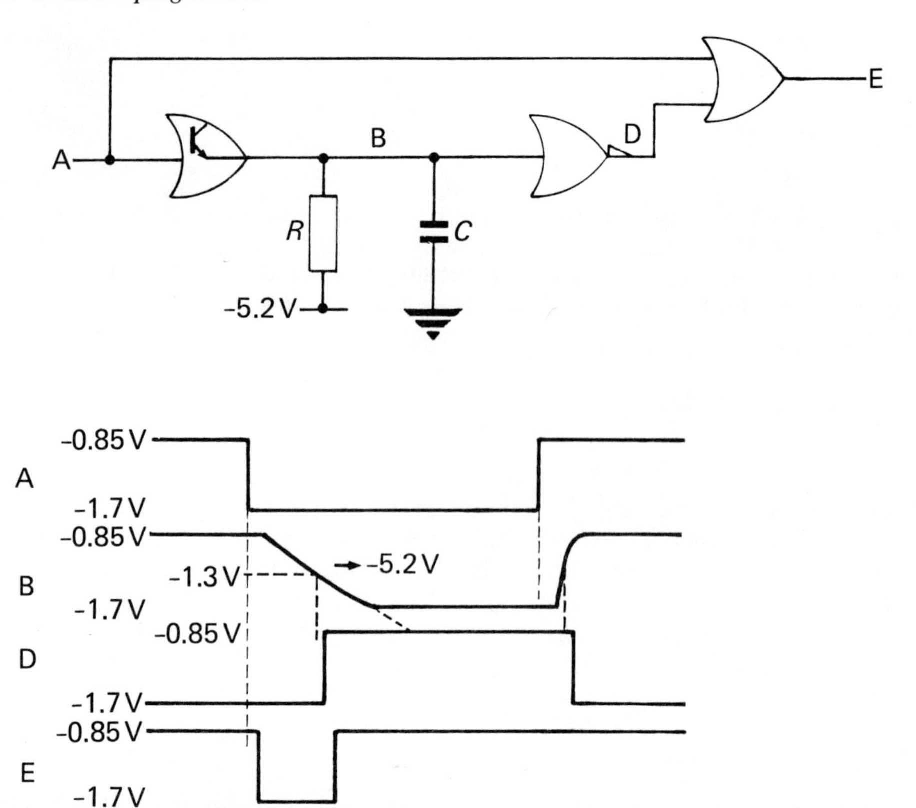

Fig. 4.5 ECL pulse generator

time constant is CR to -5.2 V (starting at -0.85 V). When the waveform B falls to -1.3 V, D switches after a further circuit delay, and the third gate produces the output pulse.

The rising edge of the waveform is fast, since the output transistor turns on, and is a low impedance source. The time constant is thus about $5C$ to 0 V. C must not be too large, or the output current will be too large. If C is 1000 pF, the rise time is around 5 ns.

Comparing the waveforms of Figs 4.2 and 4.5, it is seen that the first two gates of Fig. 4.5 form a delay to the falling input edge, but only a very short delay to the rising edge. The recovery time is thus better than that of the circuit of Fig. 4.2.

The pulse length is calculated from the exponential in the usual way. The maximum waveform fall at B is $5.2 - 0.85$ V, and the fall to the switching point of the second gate is $1.3 - 0.85$ V. Thus

$$0.45 = 4.35(1 - e^{-t/CR})$$

$$t = CR \ln \frac{4.35}{3.9}$$

$$= 0.107\, CR$$

If R is 270 Ω, and C is 1000 pF, a time of 29.4 ns is calculated. Figure 4.5 shows that two gate delays must be added to this. 29 ns is not a long time, but is more than would be obtained with a delay line.

Again, circuit tolerances make pulse width definition poor. If the gate output voltage is −0.9 V (well within specification), then $t = 0.97\,CR$, a difference of 11%. Tolerances on supply voltage level (±5% say), and on R and C, must be added to this. If full tolerances on the −0.85 V level are allowed, things are worse. It is still better than the TTL case.

Conclusion

This chapter describes a selection of useful pulse shaping circuits. There are a great many variations and alternatives. If the operation of these circuits is properly understood there should be no great difficulty in deriving the properties of others, or of designing new circuits for different situations.

Questions

1 For the circuit of Fig. 4.1, calculate values of C and R to obtain a pulse of 20 μs from a negative going edge of 4 V amplitude. R_C is 1 kΩ, V is 5 V, and the minimum β of the transistor is 40. Sketch the waveforms.

What happens if the input pulse is 10 μs wide?

2

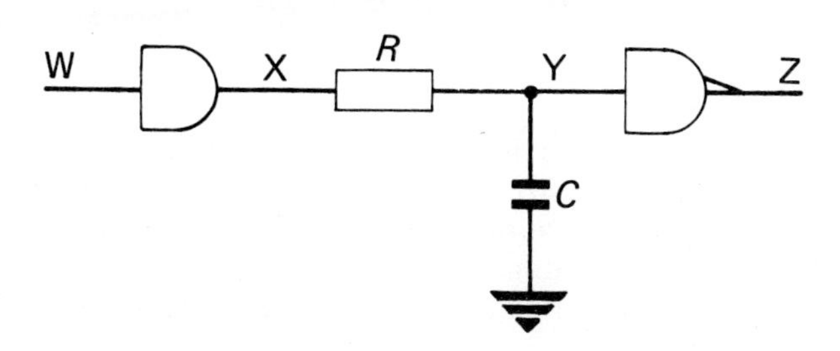

Fig. 4.6

With reference to Fig. 4.6, draw waveforms at X, Y and Z in response to a square wave applied at W. Assume TTL gates are used.

If the maximum low level input is 0.8 V and the maximum low level output is 0.4 V, what is the maximum allowable value of R? (Input resistance is 20 kΩ.)

3 Sketch the waveform at the output of each gate in the network of Fig. 4.5, assuming that RC is much less than T, the *half* period of the input square wave.

Assuming that the switching point of the transfer characteristic occurs when the input is at −1.3 V, derive the length of the pulse at the network output (gate delay = 2 ns).

Discuss what happens if T is gradually reduced to a small value relative to RC. Hence derive the rules for the proper use of the circuit in a system.

U of M (Sept 1982)

4 Derive values of R and C in question 3 for a pulse width of 200 ns (R should be ⩽1 kΩ). What is the necessary recovery time (ECL gates).

5 Sketch and describe a circuit capable of generating a pulse 200 ns wide using TTL gates. Calculate the values of all passive components (Cs and Rs) used. The maximum current *into* a TTL output is 16 mA, and the maximum current *out* of a TTL input is 1.6 mA.

Assess the tolerances on the pulse width.
What is meant by 'recovery time' in this context? Estimate its value for your circuit.
What would be the minimum pulse width available from your circuit, and why?
U of M (June 1982)

6

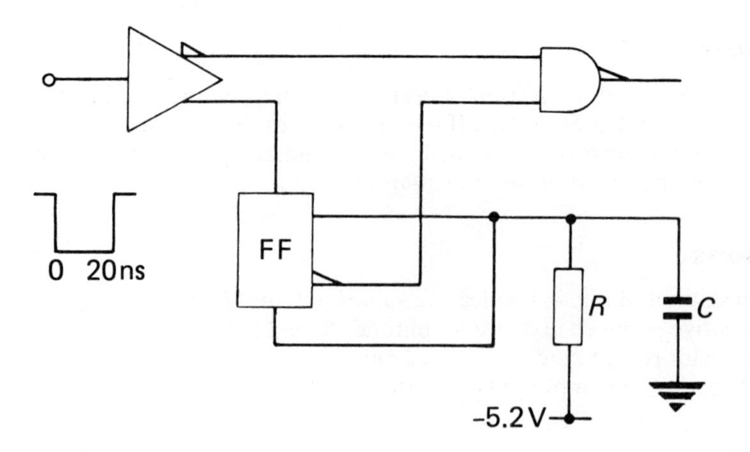

Fig. 4.7

Figure 4.7 shows the circuit of an 'active delay'. The logic levels are −0.85 V for a 'zero' and −1.7 V for a 'one'. The switching threshold is −1.3 V, and the gate delay time is 2 ns. The flip-flop (FF) operation is such that a logic 'one' applied to an input appears at the nearer output 2 ns later.

Using waveform diagrams, describe the operation of the circuit. Calculate the length of the output pulse. U of M (May 1980)

5 Monostable circuits

Introduction

The pulse shaping circuits of Chapter 4 were limited to producing outputs shorter than the input. Monostable circuits can produce outputs which are longer than the input, because they contain a storage mechanism. These circuits are the subject of this chapter.

Cross coupled amplifiers—the bistable circuit

The storage mechanism is basically that of a latch. Figure 5.1 shows two amplifiers of gain $-A$. The output of one is connected to the input of the other via a network having a gain of β, which is usually less than one. The gain of *minus A* implies that a positive going input causes the output to be negative going and vice versa. Suppose that one output, X, is high. This passes to the input of the other amplifier, the inverting action of which causes the output, Y, to be low. The low output at Y is connected to the input of the first amplifier, and confirms the high state of X. This condition is therefore stable.

Suppose that, by some means yet to be specified, the input of the first amplifier is forced slightly positive by v volts, so that it enters a region of its characteristic where A_1 is large (see, for example, Fig. 2.8). The output X now goes down by A_1v volts. The output Y will go up by $A_1\beta_2A_2v$ volts, and the change at the input to the first amplifier is $A_1\beta_2A_2\beta_1v$ volts. If this is greater than v, then an unstable region is reached, and the roles of X and Y will reverse. Changes continue until the amplifiers reach a non-linear region of their characteristic, i.e. where the value of A falls (Fig. 2.8), and the gain $A_1\beta_1A_2\beta_2$ becomes zero. Once the states have changed, the system is again stable. The gain in the stable states must be zero, since the output stays sensibly constant as the input changes. The initial disturbance must be large enough to cause the amplifiers to reach a portion of their characteristic where the loop gain, $A_1\beta_1A_2\beta_2$, is greater than unity. Under these conditions there is positive feedback. If the gain is well above unity, then the switching between stable states will be fast. Nevertheless, it is important to notice that one output changes slightly ahead of the other.

The operation described above is that usually associated with the latch. This is not

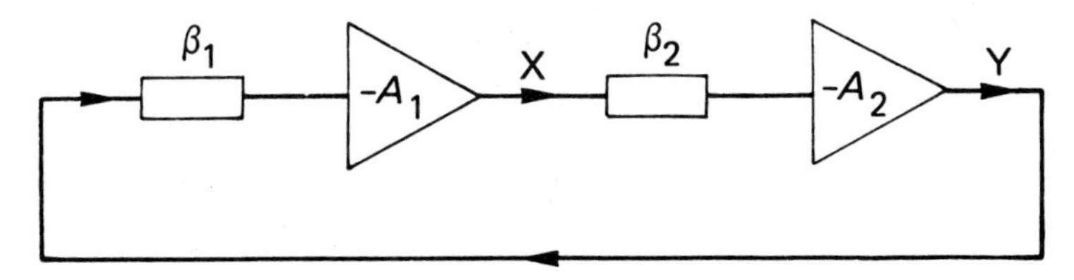

Fig. 5.1 Bistable circuit

surprising when it is realised that a gate is simply an amplifier of the comparator type, with one of the two inputs set internally. The second input to the comparator is controlled by the gate inputs to give the familiar 'gating' action. The circuit is **bistable**, because it has two stable states—X high, Y low, and vice versa.

The behaviour of the latch and flip-flop will not be discussed further here. The reader is referred to books on logic. In particular, Gibson (see Appendix A) provides a good basic description of the operation and use of logic.

Monostable circuit

Consider, now, the circuit of Fig. 5.2. The input, I, is low. One of the direct couplings of the bistable circuit has been replaced by a capacitive coupling. When any exponential changes have died away, the right hand input will be pulled positive by the resistor to $+V$. Output Y will be low, therefore, and X will be high. This state is stable.

Suppose the input goes positive sufficiently to reach the high gain region of the characteristic of the left hand amplifier. Output X goes low. Since the voltage across a capacitor cannot change instantaneously, the input Z goes lower, and output Y goes higher. If the gain of the system $A_1\beta_1$ ($\beta_2 = 1$), is sufficiently large, the input can now be removed, since the signal at Y will do the job which input I was doing.

Point X is at a constant voltage, fed from a low impedance source (e.g. a saturated transistor). Z is at a low voltage, so a current will flow in the resistor, charging up the capacitor on a time constant *CR*. When Z reaches the amplifier switching threshold, Y will go low. The output Y has a high phase which has a time period determined by *CR*. This state is *unstable*, and hence the name **monostable** circuit.

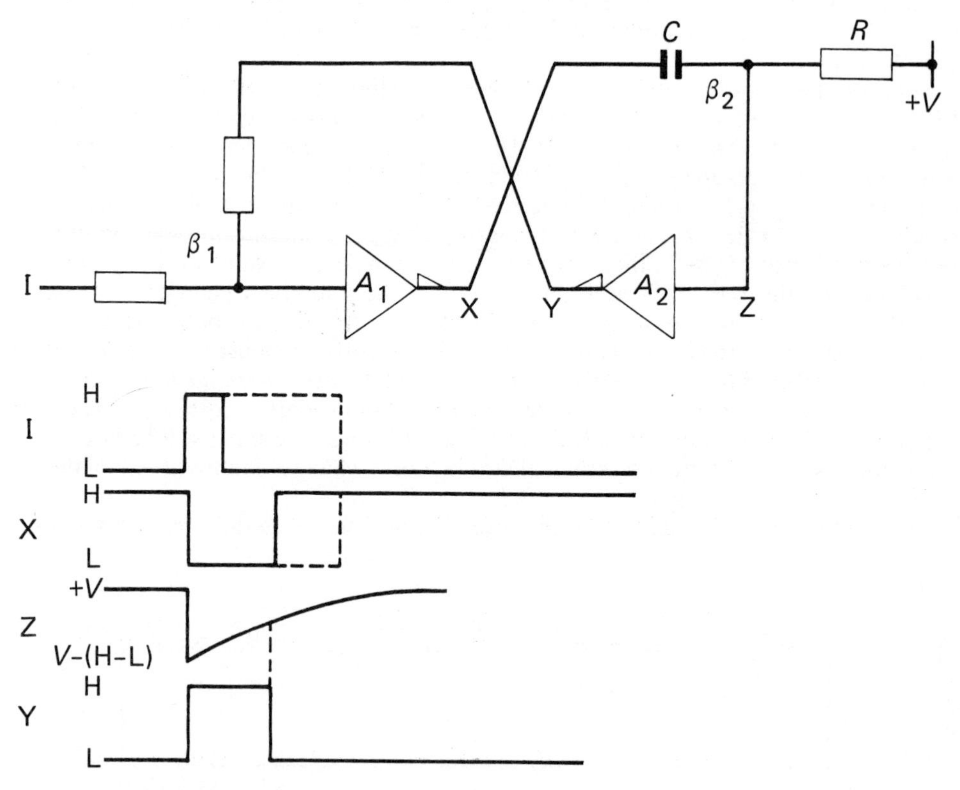

Fig. 5.2 Monostable circuit

What happens at X will depend on the state of input I. If I had gone negative prior to Y (full lines in Fig. 5.2), then X will go positive with just a circuit delay after Y goes negative (there is, of course, positive feedback). If I is still positive when the pulse at Y ends, then X will remain low until I is removed (dotted lines in Fig. 5.2).

A second input pulse occurring before the end of the timed period will not have any effect on the output, since Y has taken over the function of I.

A basic monostable circuit

Figure 5.3 shows a simple transistor circuit implementation of Fig. 5.2. Although the circuit would probably not be used as such today, the principles involved in its design are of more abiding importance. If properly understood, they will enable the reader to deduce the operation of more complex circuits.

Suppose input I to be low. Transistor T4 will be off. After all transients have died away, the base of T2 will be pulled up to approximately 0.7 V, and current will flow via resistor R. Point Y will be low. Suppose, in the first instance, that T3 is replaced by a short circuit. T1 base will be at about 0.2 V, and hence T1 is also off. Point X is therefore high (T4 is off).

Now let I go high. T4 turns on, pulling point X low. This change is transmitted by the capacitor to T2 base, which goes more negative by $(V - 0.2)$V. T2 turns off, and Y goes high, which, in turn, causes T1 to turn on, and holds X low. T4 can now be

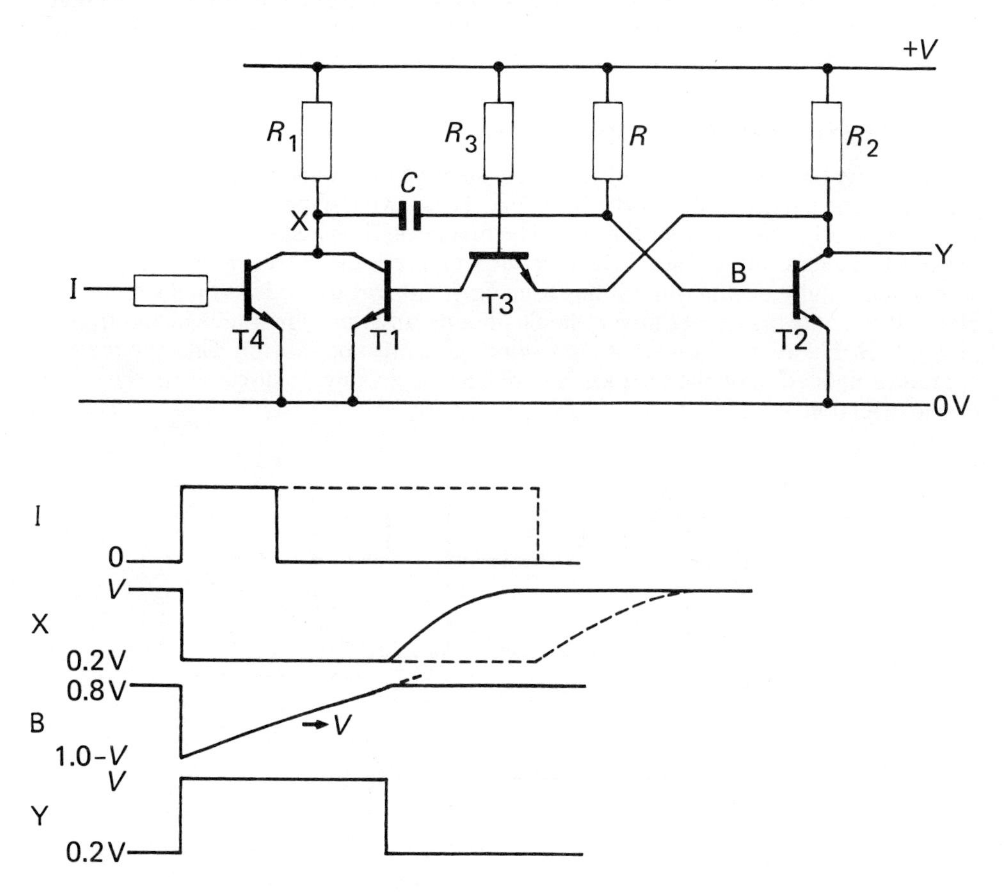

Fig. 5.3 Basic transistor monostable circuit

turned off without affecting the operation of the circuit. If T3 is a short circuit as suggested above, then Y is connected to T1 base, and cannot rise above 0.7 V due to the base–emitter diode. The presence of T3 isolates these two points, and permits Y to rise to *V*. T3 is off, but current flows through R_3 and the base–collector diode of T3 to the base of T1. In the previous condition, with Y low, T3 is saturated, and T1 base will be at 2 × 0.2 V—one due to T2, the other due to T3.

Point X is now held at a constant voltage by T1. Since no current flows into T2 base, the *C–R* network acts like an integrator. Point B rises on a time constant *CR* towards *V*. When it reaches 0.7 V, T2 switches on again. Y falls, turning off T1. If I is low, X rises, causing B to rise further, and hence T2 to come on hard. The small 'pip' in the waveform at B represents this rise in X. At this point in time, B is a fixed voltage (0.7 V), and hence there is an integrating network made up of *C* and R_1. Thus X will rise on a time constant CR_1 to *V*. Notice that this is *not* a step change as occurs in the voltage at Y. It is quite normal for this rise to be comparable in length to the pulse at Y.

Figure 5.3 also shows what happens if I is still high when T2 switches on (dotted lines). T1 is switched off, but X is held low by T4. *X and Y are not the inverse of each other*. The time constant on X as T4 turns off remains the same.

Notice the procedure used to determine the operation of this circuit. Start at some point, and make an assumption as to its state. Point B at 0.7 V is a good place. Now determine the effects of this assumption, moving round the circuit in logical order. This should lead eventually to a confirmation of the initial assumption. If it does not, then the initial assumption must be re-assessed. As an exercise, determine the stable state of the circuit, starting from the assumption that I is low and that T1 is on.

Recovery time and too rapid repetition

The recovery time of a monostable circuit is the time that must be allowed between the end of one pulse and the start of the next. To see what happens if the circuit is retriggered too soon, consider Fig. 5.4. The first pulse, I, is assumed to occur when X has reached a final value. Hence the first output pulse is the correct length. The second input pulse occurs when X has risen only part way towards its final value. Hence it can only fall by a relatively small amount, and this fall is that which is transmitted to B. Figure 5.4 shows that the second pulse must be shorter. Thus the recovery time is defined to be the time for X to reach a final value, as discussed in the previous section.

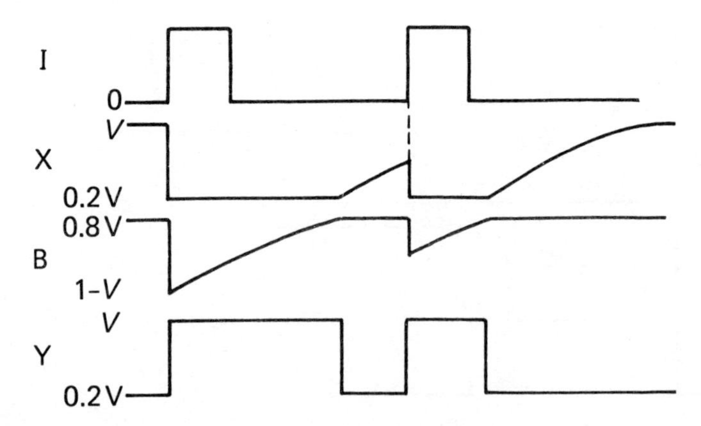

Fig. 5.4 Waveforms of Fig. 5.3 for too rapid repetition of input

Retriggerable and non-retriggerable monostable circuits

A retriggerable monostable circuit is defined as one which gives an output pulse lasting until a time T after the last input trigger. Thus, if a string of trigger pulses occur, separated by times less than T, then the monostable output will remain set as shown in Fig. 5.5(a), i.e. it is retriggered.

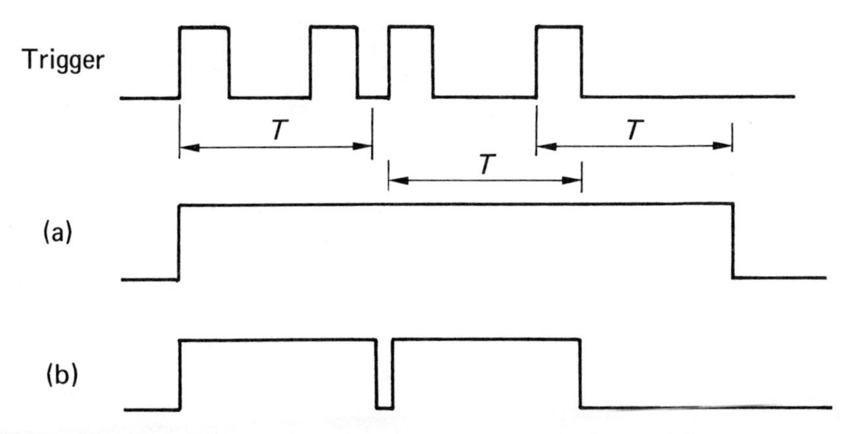

Fig. 5.5 (a) Retriggerable monostable circuit output waveforms, (b) non-retriggerable monostable circuit output waveforms

In contrast, a non-retriggerable circuit will give an output pulse of time T regardless of other trigger pulses, as shown in Fig. 5.5(b). A second output pulse is shown on the assumption that the recovery time rules have been obeyed.

The circuit shown in Fig. 5.3 is an example of a non-retriggerable circuit. The transistor T1 is saturated, so further trigger pulses clearly have no effect on the circuit. An alternative way of looking at the circuit is to regard T1 and T4 as a NOR gate.

The basis of a retriggerable circuit is shown in Fig. 5.6. Each input pulse causes the transistor to saturate. The collector voltage can never fall lower than 0 V, which is a standard level. Hence the rise is always the same. The rise cannot begin until the input pulse is removed, so a means for controlling this must be found. Figure 5.7 shows a complete circuit.

When the input I rises, the edge-triggered flip-flop Q sets, and the transistor turns on. The falling edge is exaggerated in its slowness, to show clearly what happens. When the collector voltage falls below $3V/4$, comparator 1 sets the output of the

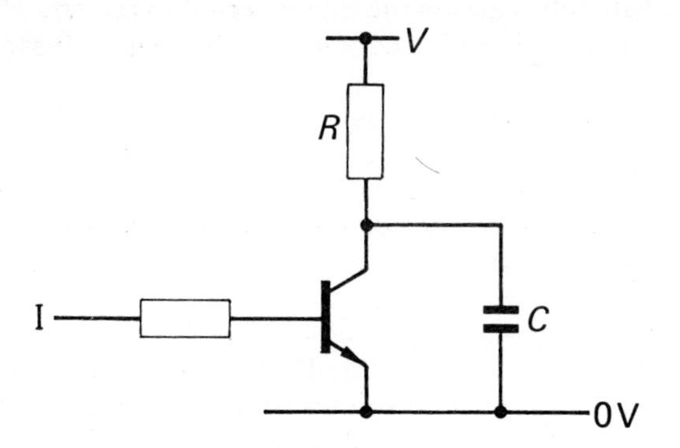

Fig. 5.6 Basic arrangement for a retriggerable monstable circuit

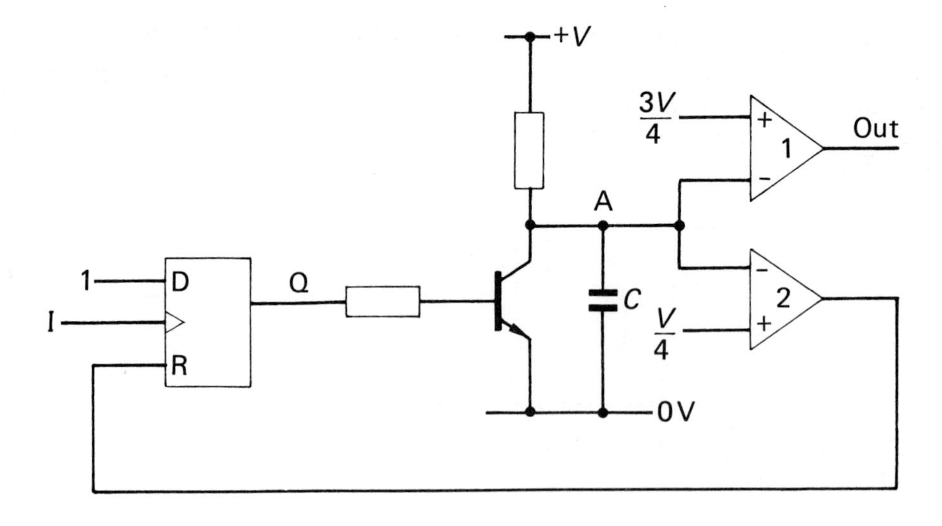

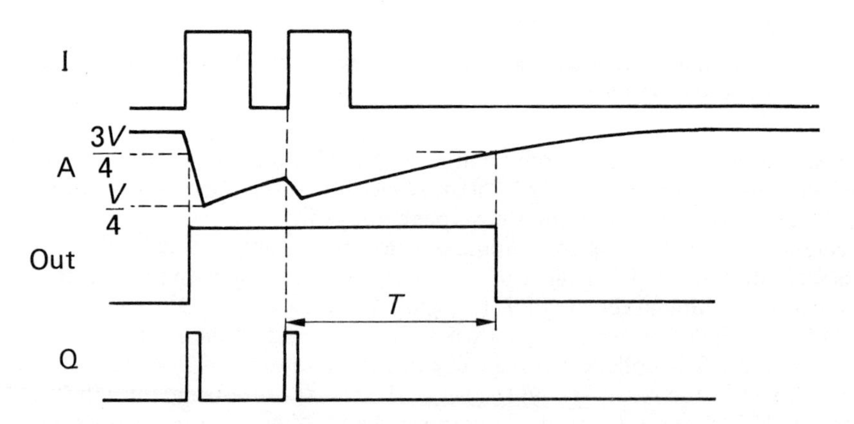

Fig. 5.7 Retriggerable monostable circuit

circuit. When the collector voltage of the transistor falls below $V/4$, comparator 2 gives an output which resets Q, allowing the timed period to begin. If a second input edge occurs, the Q again sets, causing the timed period to restart. The output falls when the transistor collector rises above $3V/4$, i.e. T seconds after the last input edge.

555 timer

The 555 timer is one of the best known and most popular integrated timing circuits available. It can be used either as a monostable circuit or as an astable circuit, and it can be obtained as a single circuit in an 8-pin package, or two circuits in a 14-pin package.

Figure 5.8 shows the timer connected as a monostable circuit. The waveforms start by assuming that the trigger signal is high, and that the discharge transistor, T, is off. The CR_T network causes the threshold voltage to rise. When it reaches $2V/3$, S causes the flip-flop Q to set, and turns T on. This discharges C. As RE is low, this situation is stable.

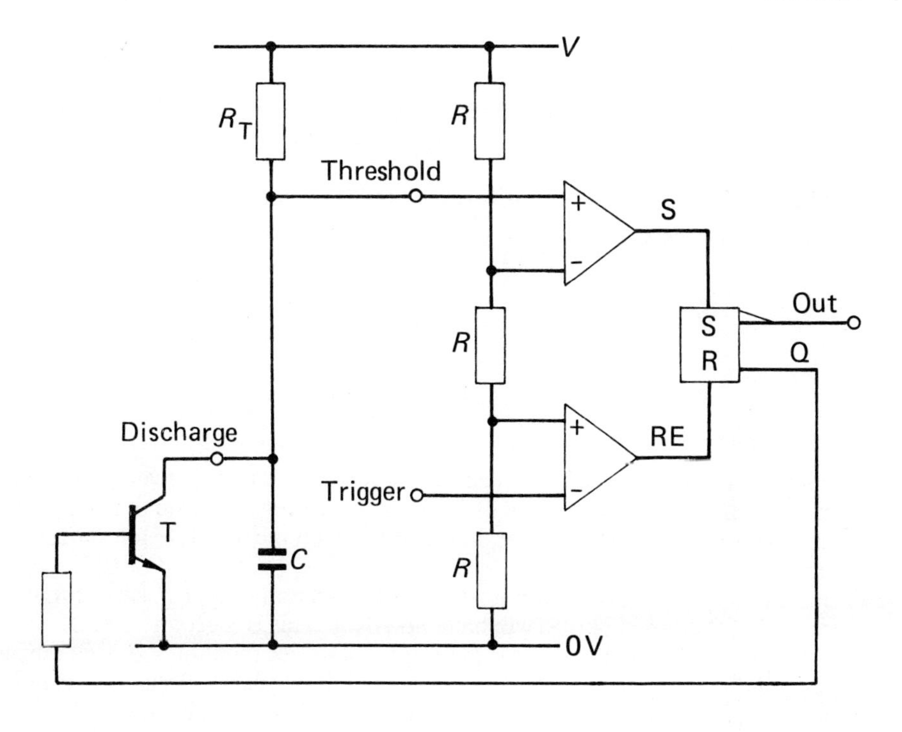

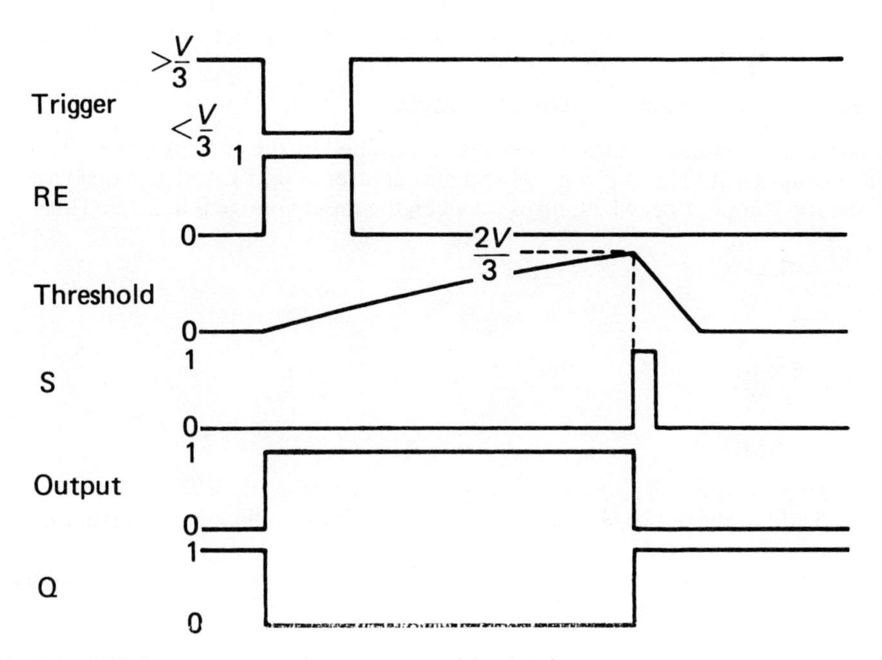

Fig. 5.8 555 timer connected as a monostable circuit

If, now, the trigger goes low, RE causes Q to reset, turning T off, and allowing the threshold voltage to rise again.

A reset input is also available (not illustrated) which can be used to terminate the timed period early. When used with the trigger input, it enables a positive edge triggered version to be designed.

The length of the timed pulse is easily calculated from the equation

$$v = V(1 - e^{-t/CR_T})$$

$$t = CR_T \ln \frac{V}{V - v}$$

$$= CR_T \ln \frac{V}{V - 2V/3}$$

$$= CR_T \ln 3$$

This is independent of V. If the tolerances in the resistors R vary between samples of the circuit, then the exact value of t will also vary. However, the ratio of resistors is usually held to within 5% between batches, even though the absolute tolerances are much worse.

Two other points are worth mentioning. Firstly, a second trigger pulse occurring before the end of the timed period will have no effect. This is therefore a non-retriggerable circuit. Secondly, if the trigger pulse is longer than the required output, the output will follow the input. Thus to operate as a monostable circuit, the required output pulse must be longer than the input pulse. To ensure that this is so, or to obtain an output pulse which is shorter than the input, a differentiating R–C network may be used between the trigger signal and trigger input. So long as the input to the 555 trigger pin remains below $V/3$ long enough to cause triggering, the circuit will work.

Maximum and minimum time periods

The maximum resistance that can be used is determined by the current required to turn on a comparator. This is given by the manufacturer as 0.25 μA. The worst case will be the top comparator, which turns on when the input voltage is at $2V/3$. Thus

$$\frac{V}{3R_T} > 0.25 \times 10^{-6}$$

and if $V = 5\,\text{V}$

$$R_T < \frac{5 \times 10^6}{0.75}$$

$$< 6.6\,\text{M}\Omega$$

With such large resistors, a good quality capacitor must be used, or the capacitor leakage will add to the 0.25 μA needed by the comparator, reducing the maximum value of R_T.

The minimum resistor is set by the current available from the discharge transistor. This is a minimum of 35 mA. Hence

$$\frac{V}{R_T} < 35$$

$$R_T > \frac{5}{35}\,\text{k}\Omega$$

$$> 140\,\Omega$$

However, if this value were used there would be no current to discharge the capacitor close to the end of discharge. Hence a value of 500 Ω minimum would seem better. The manufacturer recommends a minimum value of 5 kΩ under normal circumstances.

The largest capacitor is limited by the dissipation of the discharge transistor. The maximum discharge current is 55 mA, and the average voltage across the transistor is $V/3$. Hence, instantaneous dissipation with a 5 V supply is 92 mW. If the capacitor were 1000 μF, then the discharge time t for a 55 mA current is given by

$$55 \times 10^{-3} = 10^{-3} \frac{2V}{3t}$$

$$t = \frac{10}{3 \times 55}$$

$$= 61 \text{ ms}$$

Note that this discharge time is identical to the recovery time of this circuit.

A part of the 92 mW must be added to the package dissipation in proportion to the ratio of 65 ms to the pulse repetition period. It is necessary to hold the **duty ratio**, i.e. the ratio of 'pulse time' to 'no-pulse time', such that the recommended package dissipation is not exceeded. If, to get a low leakage capacitor, the maximum value is 10 μF, then the maximum pulse length is about 70 s. The discharge time is well under 70 ms, so the limitation is not severe.

For minimum pulse times, suppose R_T is chosen to be the manufacturer's minimum value of 5 kΩ. The minimum capacitor value should be about 100 pF so as to be much larger than any stray capacitor. The minimum pulse width is thus 500 ns. This could be reduced to about 100 ns by using an R_T of 1 kΩ. Recovery time with a 100 pF capacitor will be a maximum of 15 ns (35 mA discharge current).

A comparator monostable circuit

Figure 5.9 shows another monostable circuit. The input is connected via a differentiating network. V' is a higher voltage than V; V' might be 0 V, while V is negative. Thus, when transients have died away the output will be low.

When a negative going input takes the '−' input of the comparator below V, the output will switch, rising by H volts. The input time constant will normally be short as shown in Fig. 5.9. The '+' input of the comparator will rise by $R_2H/(R_2 + R_3)$, and will fall on a time constant $C(R_2 + R_3)$ towards V.

Let the output swing be H. The equation of the exponential is

$$v = \frac{R_2H}{R_2 + R_3} e^{-t/C(R_2 + R_3)}$$

and the final voltage when V_B reaches V' is equivalent to $V' - V$. Hence

$$t = C(R_2 + R_3) \ln\left(\frac{R_2H}{(R_2 + R_3)(V' - V)}\right)$$

Observing the waveform at point B, it will be realised that the recovery time is several times longer than the pulse length.

Conclusions

The variety of circuits available for pulse generation is enormous. This chapter is able only to describe some basic ideas, and to give a few examples. The notions of retriggerable and non-retriggerable circuits are important, as the two have quite

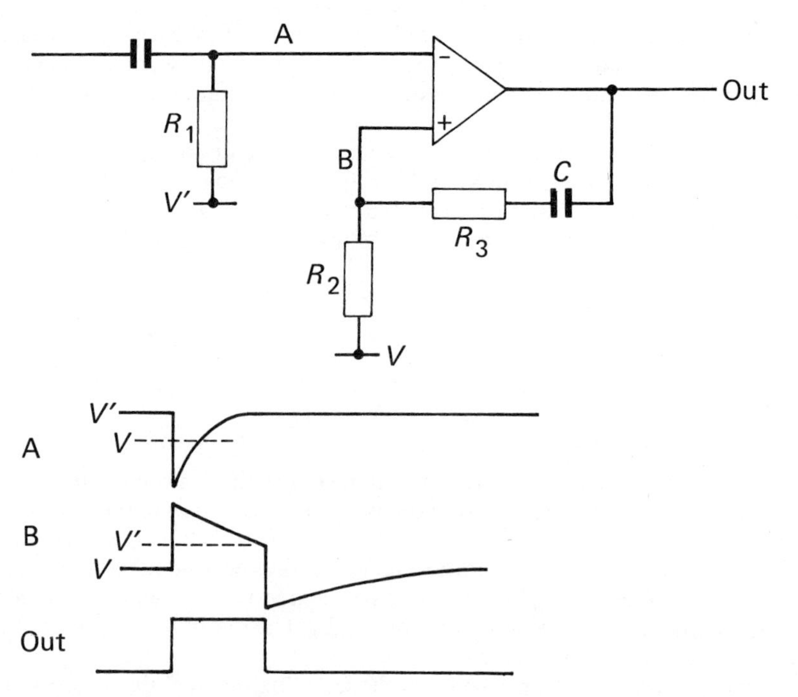

Fig. 5.9 A comparator monostable circuit

separate uses. Care must also be taken not to exceed maximum times, or to try to obtain too short pulses with inappropriate circuits. Nor must recovery time be ignored.

The chapter specifically does not attempt to describe the TTL monostable circuits 74121, 122, 123, etc.. The internal operation is quite complex in detail, and is by no means easy to understand. The circuits are useful, however, and should not be forgotten.

Questions

1 In the circuit of Fig. 5.6, I has amplitude 3.7 V, V is 5 V, $R = 470\,\Omega$, $R_b = 3\,k\Omega$, $C = 1\,nF$ and the transistor $\beta = 40$. Calculate the rise and fall times of the output.

2 The trigger of Fig. 5.8 is at 3.5 V, and the threshold is at 2 V. Determine the steady state of the circuit ($V = 5\,V$).

Sketch waveforms of the circuit after I goes low for a short time. What happens if I remains low?

Calculate values for R_T and C for an output pulse of 1 second.

3 With the aid of waveform diagrams, distinguish between retriggerable and non-retriggerable monostable action.

Sketch the circuit diagram of a monostable circuit and describe its operation (comparators and gates may be regarded as circuit components). State whether your circuit is retriggerable or not, saying why.

Estimate the recovery time and maximum and minimum pulse widths, stating any assumptions you make. U of M (June 1983)

Repeat question 3 for several of the circuits given in the text.

4 Suggest values of R_T and C for pulse widths of 200 ns and 2 s, using the circuit of Fig. 5.8. What would be the recovery times? (Assume 35 mA available from T, $V = 15\,V$.)

6 Astable circuits or oscillators

Introduction

A monostable circuit is produced by replacing one of the β networks of Fig. 5.1 by a differentiating network, as shown in Fig. 5.2. If both β networks are replaced thus, the circuit will oscillate. This is because when the output of Fig. 5.2 falls, it starts a timing period for the other half of the circuit, which ends by retriggering the first half. This circuit is little used today, but forms the basis of all circuits. This chapter discusses various methods of designing astable circuits.

555 timer as an astable circuit

Figure 6.1 shows the 555 timer connected as an astable circuit. To determine the mode of operation, start by making an assumption. Let T be on, so that X is at 0 V. If Y is not at 0 V, it must be falling on a time constant CR_2 towards 0 V. When it falls below $2V/3$, S goes low, but Q is not affected. When Y falls below $V/3$, RE causes Q to be reset, and T turns off.

Once T has turned off, Y begins to rise, but now on a time constant $C(R_1 + R_2)$, towards V. The voltage at point X will jump suddenly. As Y starts at $V/3$, X will jump to

$$\frac{V}{3} + \frac{R_2}{R_1 + R_2}\frac{2V}{3}$$

and Y rises above $V/3$ rapidly, causing RE to go low. When Y reaches $2V/3$, S causes Q to set, T to turn on, and the cycle repeats. When Y is at $2V/3$, X is at a voltage of

$$\frac{2V}{3} + \frac{R_2}{R_1 + R_2}\frac{V}{3}$$

In calculating the time period 'V' is $2V/3$, and v is $V/3$. The time constants are CR_2 to 0 V, and $C(R_1 + R_2)$ to V. Times are therefore

$$t_1 = CR_2 \ln 2$$

and $$t_2 = C(R_1 + R_2) \ln 2$$

These are independent of V. However, it is obviously impossible to obtain equal high and low portions of the waveform.

The maximum value of $R_1 + R_2$ is determined by the 0.25 μA current required by the S comparator, as it was for the monostable circuit. When using very large resistors, the voltage at X will be affected by the 0.25μA, which was taken as negligible in the above calculations.

A second astable circuit

What is required of an oscillator circuit is a positive feedback system during switching, and a negative feedback system during the 'stable' part of the waveform to give a gain

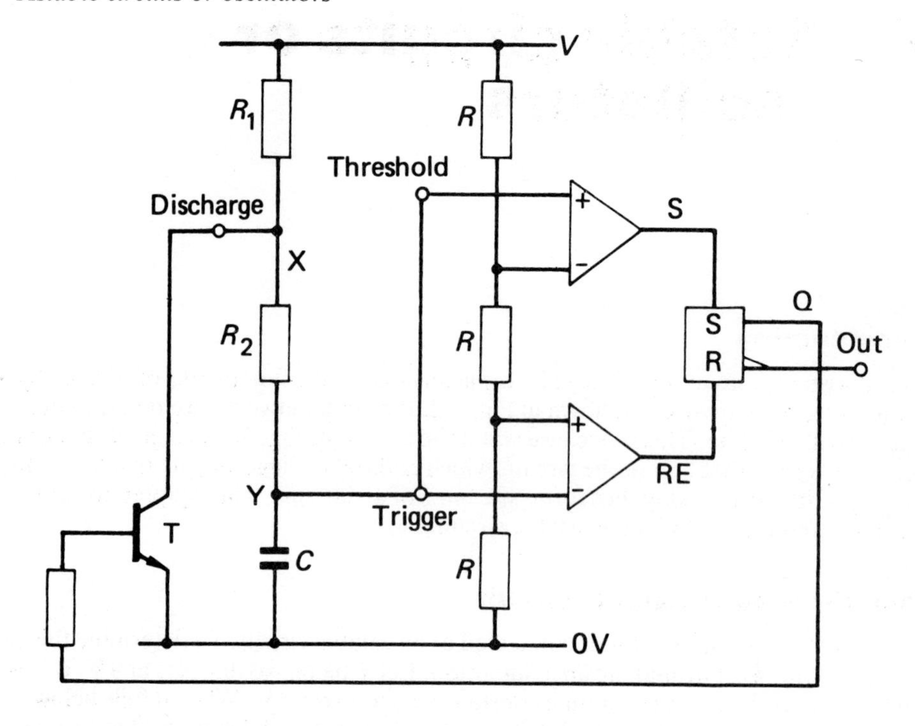

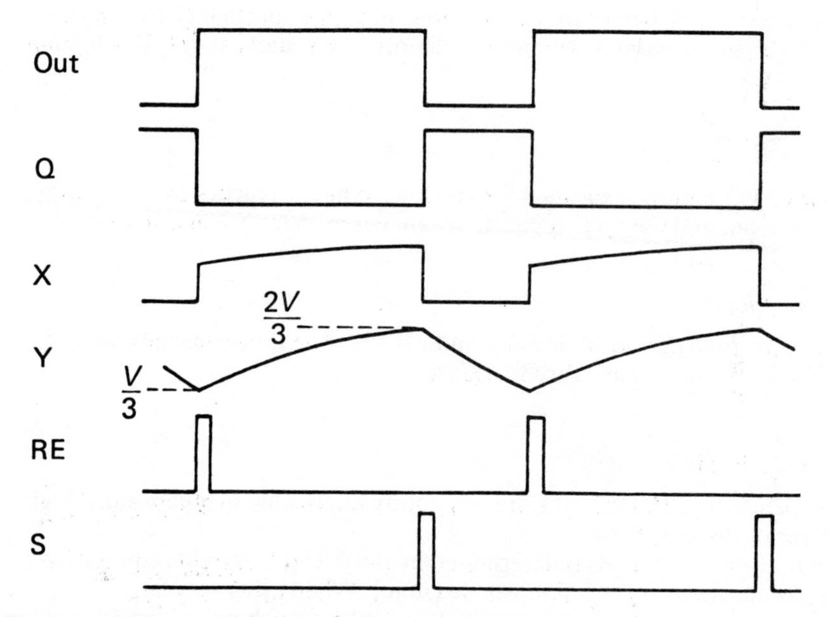

Fig. 6.1 555 timer connected as an astable circuit

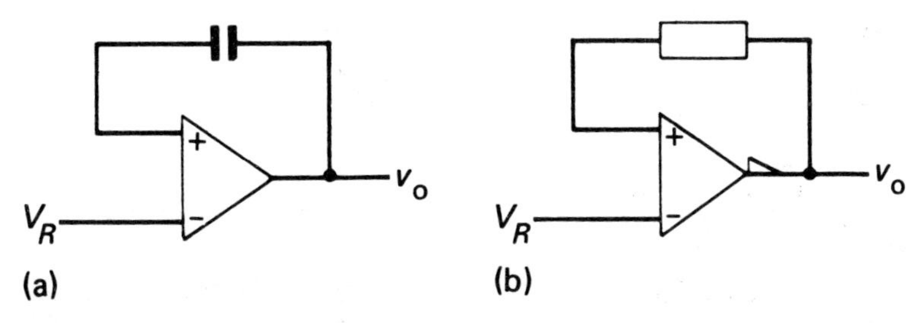

Fig. 6.2 Components of an astable circuit

of unity. It is also necessary to have a C–R network, the condition of which is changed during switching.

The positive feedback during switching can be obtained by connecting a capacitor between the input and output of an amplifier of positive gain, as shown in Fig. 6.2(a). Any change at the input is amplified and, as the voltage across a capacitor cannot change instantaneously, this is fed directly back to the input.

Unity gain is obtained by use of resistive feedback round a circuit of negative gain, as shown in Fig. 6.2(b). If the whole of the output is fed back to the input ($\beta = 1$), and the gain A is large, then

$$v_o = \frac{-AV_R}{1+A}$$

$$\simeq -V_R$$

i.e. a gain of unity.

Now consider the input of a single amplifier to be connected to one end of both a capacitor and a resistor. The other end of the capacitor is connected to the output of a positive gain amplifier, and the other end of the resistor to the inverse of this. Thus the two remote ends will move in opposite directions. Figure 6.3 shows the circuit and waveforms.

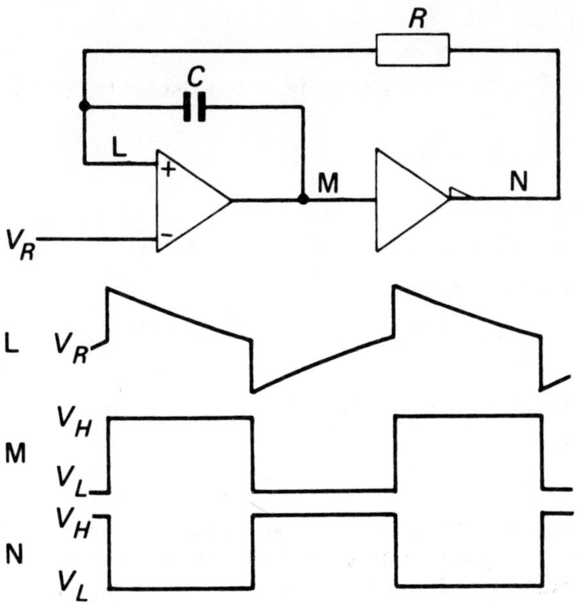

Fig. 6.3 Comparator oscillator

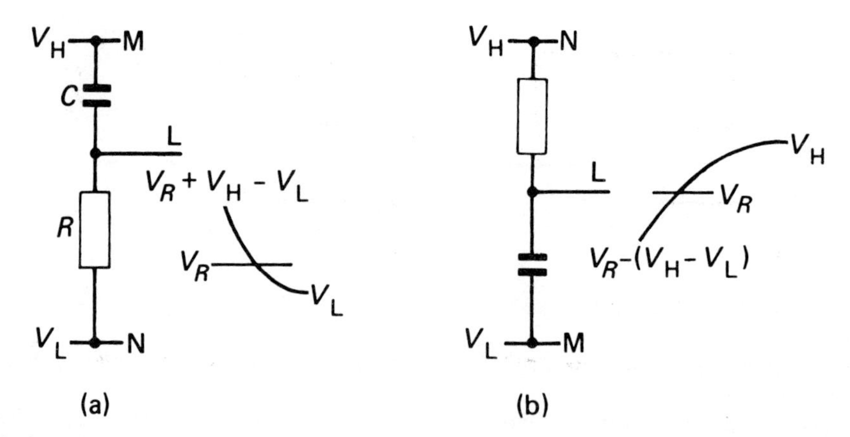

Fig. 6.4 *R–C* states for Fig. 6.3

Suppose point L to be just slightly above V_R. Then M rises sharply due to the amplifier gain, and this is communicated to L. The voltage at N falls. Figure 6.4(a) shows the situation as it affects the *C–R* network. In this case, point L will fall on a time constant *CR* towards V_L, the lower of the amplifier voltage levels.

When L falls to V_R the amplifier output M will fall a little, and this is fed by the capacitor to L, giving a large and fast fall at M due to the positive feedback. Point N will rise to V_H. Figure 6.4(b) illustrates the conditions. L will now rise on a time constant *CR* towards V_H.

Consider the waveform of L. Initially, the voltage is $V_R + V_H - V_L$, and falls towards V_L. The amplitude of the swing is thus $V_R + V_H - 2V_L$, and the equation is

$$v = V_L + (V_R + V_H - 2V_L)e^{-t_1/CR}$$

(to check, set $t = 0$ and $t = \infty$). The time period ends when $v = V_R$. Hence

$$V_R = V_L + (V_R + V_H - 2V_L)e^{-t_1/CR}$$

$$t_1 = CR \ln \frac{V_R + V_H - 2V_L}{V_R - V_L}$$

For the other period, when L is rising, the initial voltage is $V_R - (V_H - V_L)$, and the amplitude is thus

$$V_H - V_R + V_H - V_L = 2V_H - V_R - V_L$$

The equation of the curve is

$$v = V_R - V_H - V_L + (2V_H - V_R - V_L)(1 - e^{-t_2/CR})$$

Switching occurs when $v = V_R$. Hence

$$V_R - V_R - V_H - V_L = (2V_H - V_R - V_L)(1 - e^{-t_2/CR})$$

$$t_2 = CR \ln \frac{2V_H - V_R - V_L}{2V_H - V_R - V_L - V_H + V_L}$$

$$= CR \ln \frac{2V_H - V_R - V_L}{V_H - V_R}$$

If V_R is set half way between the two voltage levels, then $t_1 = t_2 = CR \ln 3$. The important factor here is that this is independent of the logic levels.

An interesting example of this is an oscillator built from gates in ECL, as illustrated in Fig. 6.5. The ECL circuit is a comparator with V_R set internally to -1.3 V. The logic

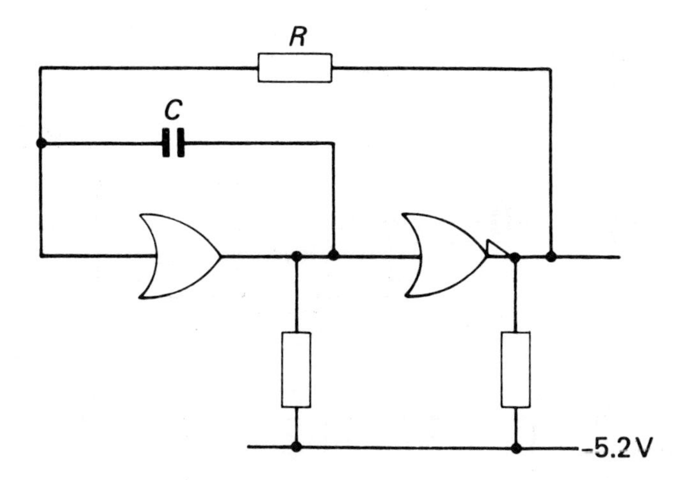

Fig. 6.5 ECL oscillator

levels are −0.85 V and −1.7 V. V_R is thus half way between the two levels, allowing for tolerances. The output impedance of the gates is low (less than 10 Ω), so that the necessary pull-down resistors at the output have very little effect on the timing. The frequency of oscillation is independent of the logic levels, provided that the switching threshold is half way between them. This will be true to within close limits, since the ratio of resistors internal to an IC is constant to quite good tolerance (even though absolute tolerance is poor). It will be noticed that in discussing previous ECL circuits, timing tolerances were poor due to voltage level variations between samples. This circuit overcomes the problem, though only for oscillator circuits.

Another astable circuit

Figure 6.6 shows another astable circuit. Let the ratio $R_3/(R_2+R_3)=\beta$, and the output voltages be H and L. The two values of the voltage at the '+' input of the comparator will be βH and βL. Let the output be H. The '−' input of the comparator will be rising. When it reaches βH the output will fall to L, and the '+' input of the comparator falls to βL. The '−' input of the comparator falls on a time constant CR_1 towards L until it reaches βL, when the circuit switches once more. The half period is calculated from

$$\beta(\mathrm{H}-\mathrm{L}) = (\mathrm{H}-\beta\mathrm{L})(1-\mathrm{e}^{-t/CR_1})$$

$$t_1 = CR_1 \ln \frac{\mathrm{H}-\beta\mathrm{L}}{\mathrm{H}-\beta\mathrm{L}-\beta\mathrm{H}+\beta\mathrm{L}}$$

and

$$t_2 = CR_1 \ln \frac{\beta\mathrm{H}-\mathrm{L}}{\beta\mathrm{H}-\mathrm{L}-\beta\mathrm{H}+\beta\mathrm{L}}$$

hence the period is

$$t = CR_1 \ln \frac{\mathrm{H}-\beta\mathrm{L}}{\mathrm{H}(1-\beta)} \frac{\beta\mathrm{H}-\mathrm{L}}{\mathrm{L}(\beta-1)}$$

With this circuit, L must be negative and H positive.

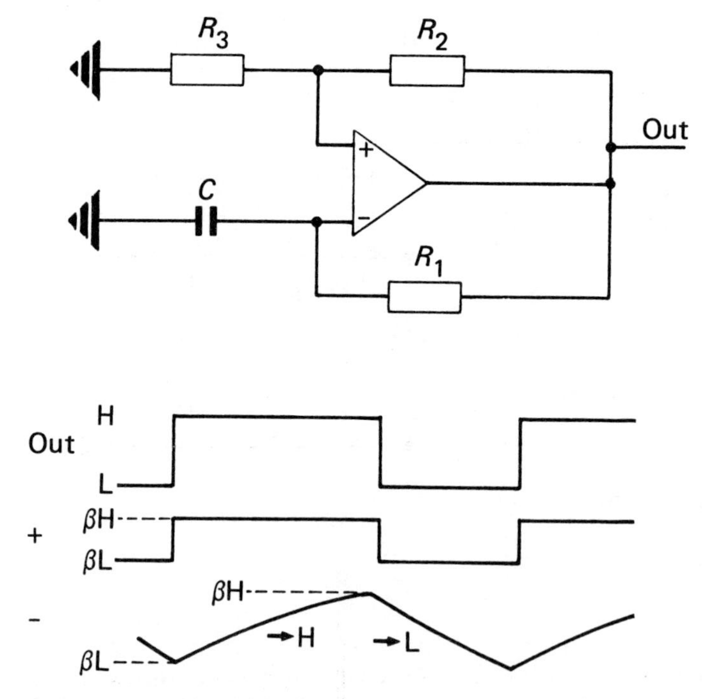

Fig. 6.6 Another comparator astable circuit

Crystal-controlled oscillators

Certain types of mineral crystal exhibit electromechanical effects. In particular, when cut in an appropriate manner, a quartz crystal acts like a series tuned circuit as shown in Fig. 6.7. Analysis of this circuit shows that at a frequency f_0 given by

$$f_0 = \frac{1}{2\pi\sqrt{LC}}$$

the inductive and capacitive effects cancel, and the circuit looks like a pure resistance. Further, if an electrical 'shock' is applied, energy will oscillate between the inductance (in magnetic form) and the capacitance (in electrical form). In the quartz the oscillation is acoustic. The frequency of oscillation is f_0, and is well defined and highly stable. Oscillators using such crystals are therefore used where stability of frequency is required. In particular, two oscillators of similar design at locations remote from each other will oscillate at frequencies very close together.

When a series tuned circuit is driven by an edge from a voltage source, an oscillation at the natural frequency of the circuit results, dying away after a number of cycles; the number is higher the higher the Quality, Q, of the circuit. Q is defined as $2\pi f_0 L/R$. For quartz crystals with an f_0 of 4 MHz the values in the circuit might be as follows:

$$L = 0.1\,\text{H} \quad C = 0.02\,\text{pF} \quad R = 500\,\Omega \quad Q = 5000$$

Figure 6.7 shows one form of crystal controlled oscillator. In this circuit, R_1 and R_2 act as bias resistors. Suppose that the output has just gone positive. The tuned circuit causes a half cycle of sine wave to occur at the '+' input of the comparator. When it falls to the bias level the output falls. This introduces a further 'shock' to the tuned

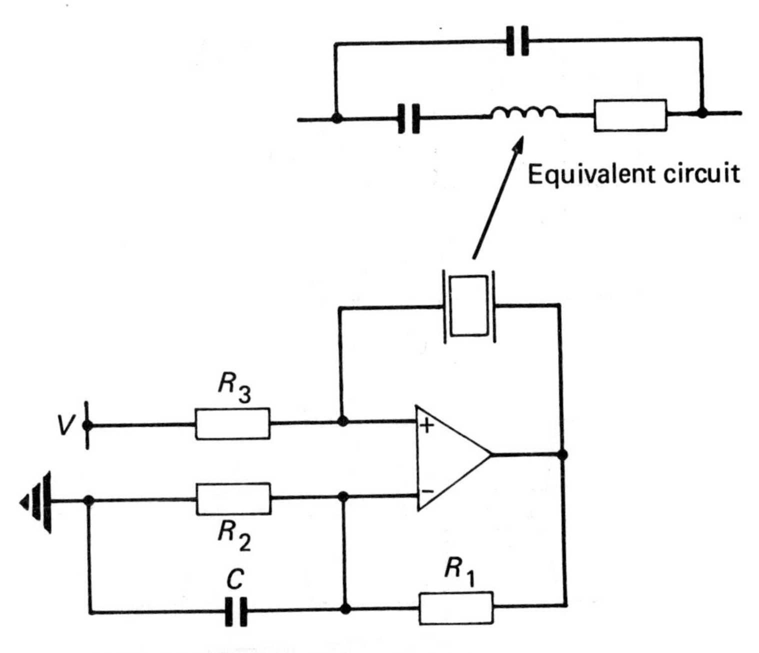

Fig. 6.7 A comparator crystal-controlled oscillator

circuit, which executes a half cycle in the negative direction. The time of switching is thus determined solely by the natural frequency of the crystal. The capacitor across R_2 is chosen to decouple the bias network at the frequency of oscillation. It is recommended that $1/\omega C$ be less than $R_1/500$.

Figure 6.8 shows a crystal controlled oscillator much used in microprocessors. The amplifiers are, in fact, TTL inverting circuits (e.g., 74LS04). The waveforms are those measured on the author's own microcomputer for a 4 MHz oscillator.

The circuit acts as a pair of amplifiers in the linear region, amplifying the input current to give a voltage at the output. The coupling capacitor is to allow the first amplifier to have its 'average' input set; some circuits do not allow this. A small capacitance on pin 3 of the diagram would provide attenuation of signals at multiples of the crystal frequency (harmonics), and ensures that oscillation takes place at the wanted frequency.

Emitter coupled astable circuit

Figure 6.8 shows the basic arrangement of an emitter coupled astable circuit. Suppose, initially, that T1 is off and T2 is on. In this case $V_{c1} = V$, $V_{c2} = V - 2IR$, $V_{b2} = V$ and hence $V_{e2} = V - 0.7$ V. Now V_{e1} will be positive with respect to $V - 2IR - 0.7$ V, since T1 is off. As T1 is off, and T2 emitter is a voltage source, V_{e2} will fall linearly at a rate given by $dv/dt = I/C$. Eventually the emitter voltage of T1 falls below $V - 2IR - 0.7$ V, and T1 begins to turn on. The collector of T1 starts to fall, which causes T2 to start to turn off. As a result, the collector of T2 begins to rise, turning T1 on harder, and, incidentally pulling both the base and emitter of T1 up. As the emitter of T1 rises, so must the emitter of T2, since the voltage across a capacitor cannot change instantaneously. The roles of T1 and T2 have now reversed from that at the start of the description. Figure 6.9 shows the waveforms.

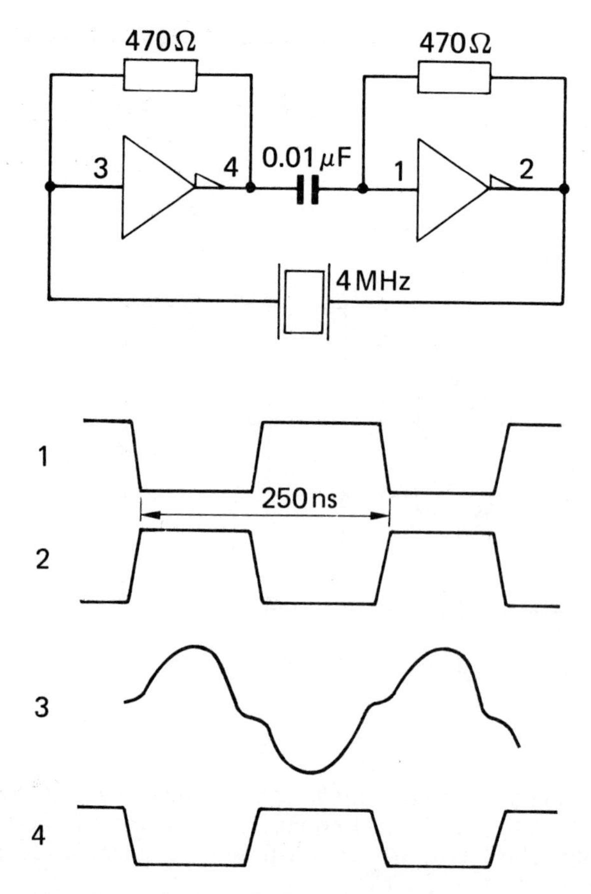

Fig. 6.8 Microprocessor crystal-controlled oscillator

With the circuit as drawn in Fig. 6.9, the transistors must not be allowed to saturate with a current of I. If this could happen then, on switching on, a current of I might flow in each transistor, saturating both, and the oscillation would not start. Thus $IR < 0.5$ V, say. It may also be required to make $2IR < 0.5$ V if saturation during oscillation is to be avoided as well. To get a larger output swing, an emitter follower may be inserted between a collector and the associated base of the other transistor. A diode may then be connected across the collector resistor to avoid saturation (see Fig. 6.10). This would allow the value of I to be varied without varying the output voltage swing. By varying I it is then possible to construct a variable frequency oscillator.

Figure 6.10 shows the essentials of a voltage controlled oscillator. An emitter coupled oscillator makes use of two fixed current sources from T3 and T4. T5 and T6 have a variable voltage between the bases. When one base goes positive, the other goes negative by the same amount. The current I_C is thus shared between T5 and T6. Current in T5 will split equally between the two collectors, since they have a common geometry. It follows that I of the oscillator will be varied by the variable input voltage, and hence the frequency of oscillation will also vary. Any non-equality in the current splitting of T5 will appear as an inequality in the two half periods of the cycle, and not in a different frequency.

By replacing the capacitor by a crystal, the oscillator can be made to run at a very stable frequency. The waveform V_{e2} of Fig. 6.9 would become a half sine wave, T2 turning on again as the voltage falls sufficiently low.

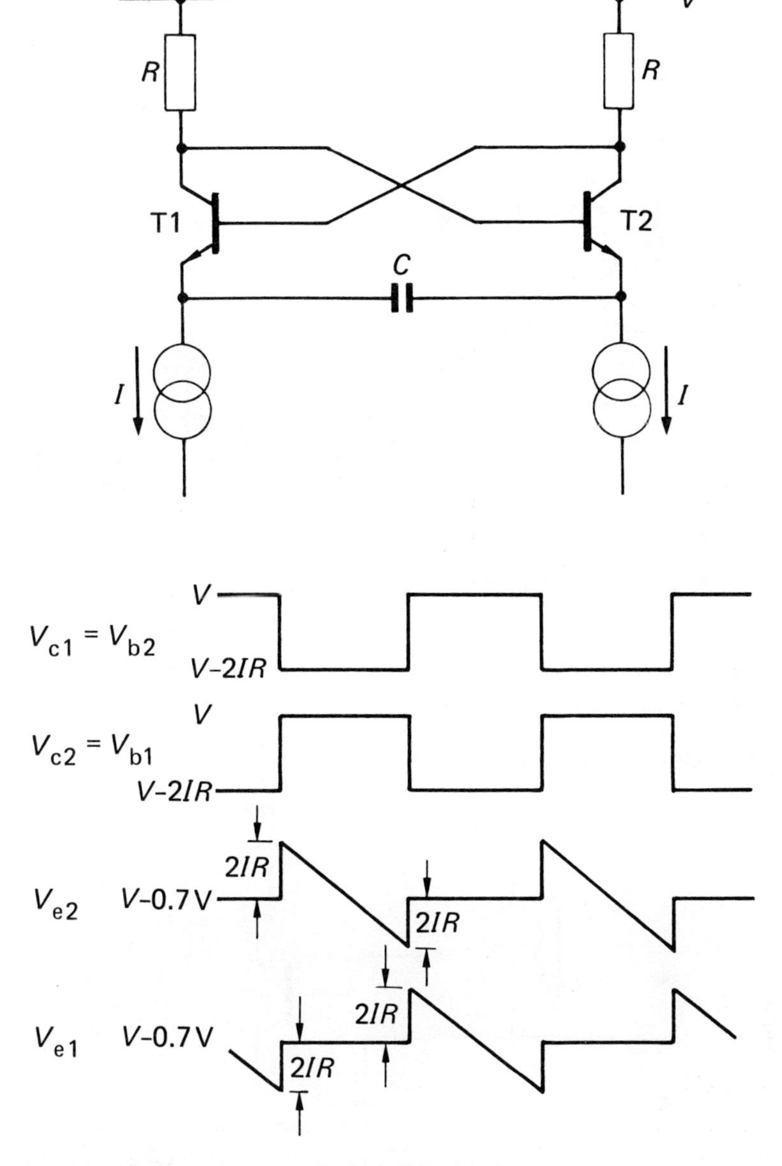

Fig. 6.9 Principle of the emitter coupled astable circuit

Another form of voltage controlled oscillator

Figure 6.11 shows another form of oscillator. Suppose that initially the state of the Schmitt trigger is such that transistor T3 is off. No current can flow in either T1 or T2, so I must flow into C, causing it to charge up. Eventually the switching threshold of the Schmitt trigger is reached, and T3 is turned on. T1 and T2 now conduct. The base of T1 falls to a low voltage (about 0.9 V above the emitter of T3). The top of D1 is thus pulled down, causing D2 to turn off. Hence the whole of I flows in T1. The arrangement of T1 and T2 is a well known method of making a current source (known as a **current mirror**) in which the collector currents of the two transistors are equal. Hence

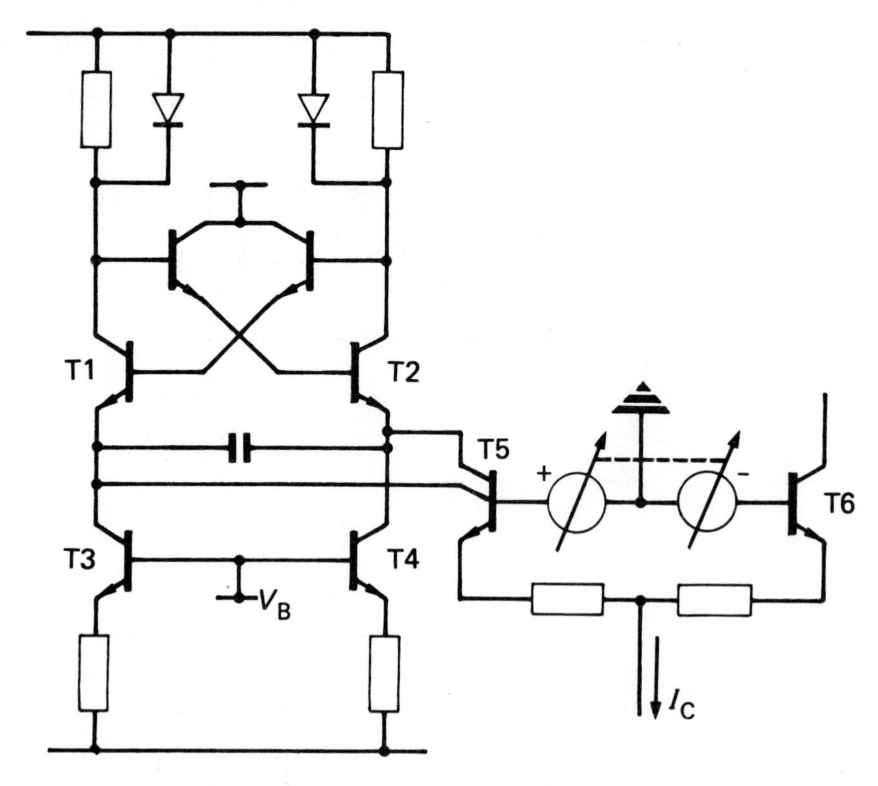

Fig. 6.10 Voltage controlled oscillator

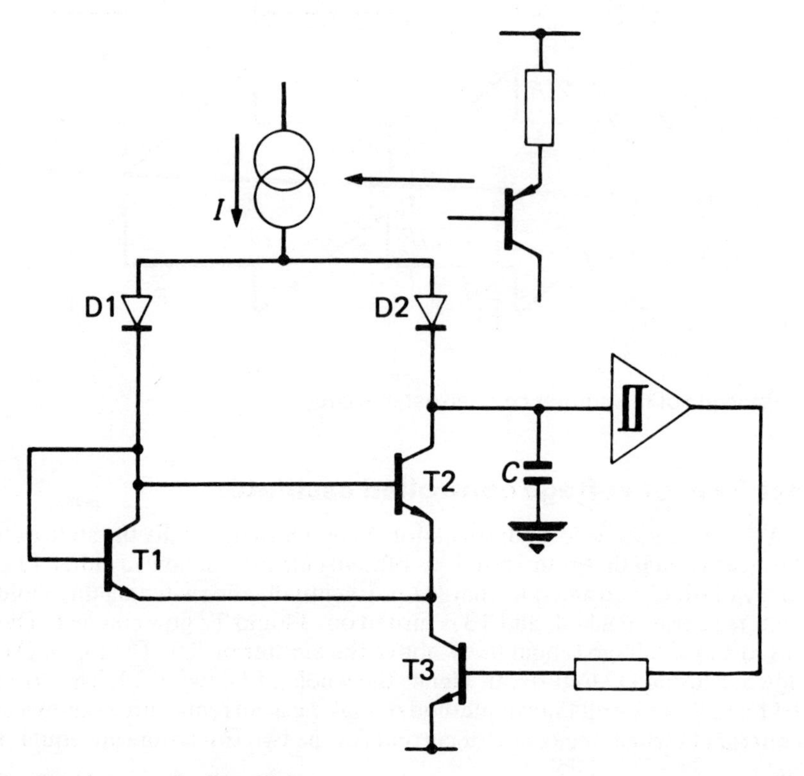

Fig. 6.11 Ramp controlled oscillator

a current equal in magnitude to I flows in T2, and therefore out of C, discharging C at the same rate as I charged it when the Schmitt trigger was in the other state. This discharge will therefore take the same time as the charge, and the Schmitt trigger will switch back to its original state, generating a square wave output. If the current I is generated, in essence by a p–n–p transistor (as shown inset), then by controlling the base voltage externally, the frequency can be varied.

Conclusions

As in the case of monostable circuits, the circuits described here are a very limited subset of those possible, although it is hoped that they are reasonably representative. No attempt is made to assess maximum and minimum frequencies, but the estimation can be approached in the same way as that used to calculate maximum and minimum pulse widths of the 555 monostable circuit. Great stability of frequency can be obtained by the use of crystals.

Voltage controlled oscillators are used for a variety of purposes, including digital voltmeters, and in the matching of the oscillator frequency to that of a 'self timed' data stream. This last application will be described in Chapter 9.

Questions

1 Design an astable circuit capable of producing an output square wave of frequency 250 kHz at TTL voltage levels. A single power supply rail is available.

Describe the operation of your circuit, and sketch all relevant waveforms (internal operation of a proprietary circuit would be expected, if chosen).

Comment on the merits or demerits of your circuit. U of M (May 1980)

2 For the circuit of question 1, what are the maximum and minimum frequencies?

3 Design an ECL oscillator to operate at 100 Mhz.

4 Design an emitter coupled astable circuit to run at 1 MHz, and with an output swing of at least 1 V.

5

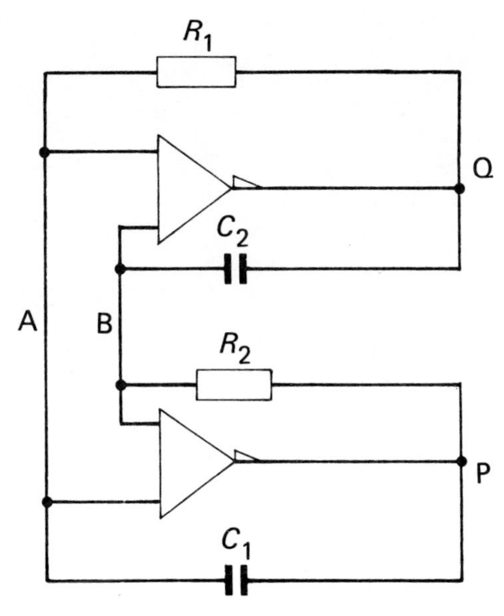

Fig. 6.12

Assuming P and Q in the circuit in Fig. 6.12 are in opposite logical states at time zero, explain the operation of the circuit. Logical levels are 0 V and 3.5 V.

Draw waveform diagrams for each of the points A, B, P and Q.

Deduce all time periods involved in terms of R_1, R_2, C_1, C_2 and logic levels.

What conditions must apply in order that both comparators switch at 1.75 V? What advantage follows from this?
U of M (June 1982)

6 Explain the operation of the circuit of Fig. 6.11. Sketch waveforms at the terminals of the Schmitt trigger circuit.
U of M (June 1983)

7 Transmission lines

Introduction

Strictly speaking, any connection between two points is a transmission line. A signal passing over the link will travel at the velocity of an electromagnetic wave in the material of the line. The best known electromagnetic wave is light. This travels at a velocity of $3 \times 10^8\,\mathrm{m\,s^{-1}}$ in a vacuum. It can be shown that

$$\text{velocity of electromagnetic wave} = \frac{1}{\sqrt{\varepsilon\mu}}$$

where ε and μ are characteristics of the material of the line, known respectively as the permittivity and permeability. For a vacuum, these values are designated ε_0 and μ_0. For other media, relative values ε_r and μ_r are defined.

It might be thought that the velocities concerned would be such that the effects could be ignored. However, digital circuits are now available which are capable of operating in well under 1 ns. In this time light travels approximately 330 mm in vacuo, and between 150 and 200 mm in most real lines. On this basis it is clear that the time of transmission cannot be ignored.

Types of transmission line

If the transmission line is not to cause serious problems, its characteristics must be determined, and they must be reasonably uniform over the length of the line. Figure 7.1(a) is a coaxial cable, familiar as the connection to the aerial of a television set. It consists of an inner conductor, surrounded by an outer cylindrical one, which is earthed. The two are separated by an insulating material. With this structure, the electric and magnetic fields, which are characteristic of electromagnetic waves, are uniform over the length of the line, and predictable over its cross section.

Figure 7.1(b) shows a twisted pair cable. One of the wires is earthed. The characteristics are determined by the separation of the two wires, and by the pitch of the twist.

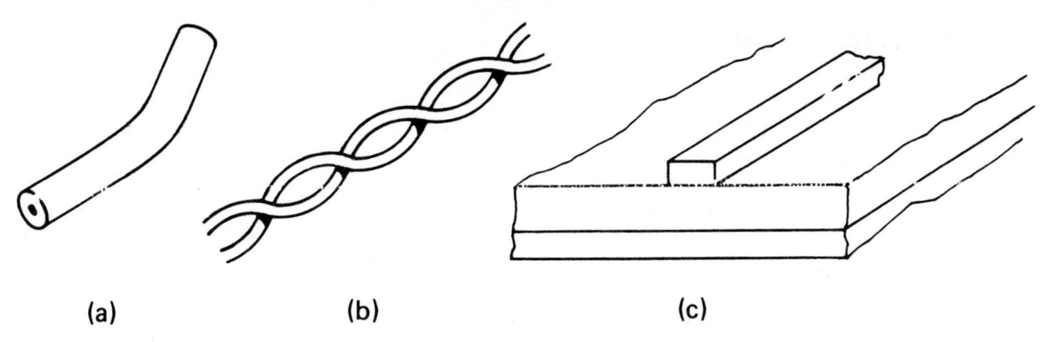

Fig. 7.1 Types of transmission line: (a) coaxial cable, (b) twisted pair, (c) strip-line

The third type of line is that found on printed circuit boards. Here a wire is run over an earth plane, and separated from it by an insulator. The characteristics depend on the dimensions of the conductor, and its separation from the earth plane (see Fig. 7.1(c)).

Basic principles

Consider the element of line run parallel to an earth plane as shown in Fig. 7.2. A current flowing through the wire generates a magnetic field round the wire according to the right hand corkscrew rule. Further, since there is a voltage on the wire, there will be an electric field between the wire and earth. The magnetic effect can be represented by an inductance, and the electric effect by a capacitance. Thus the line can be represented as in Fig. 7.3, where L and C are the inductance and capacitance *per unit length* respectively. As $\delta x \rightarrow 0$, this representation tends to a true representation of a line. The full representation should also include the series resistance of the wire and the parallel conductance of the insulation. For the purposes of this text these are sufficiently close to zero to be ignored.

Now consider a voltage suddenly applied to one end of a transmission line. Since it takes some time for a signal to travel down the line, the current flowing will initially be independent of the *termination* at the far end. Thus it is possible to write

$$V = IZ_0$$

where Z_0 is the **characteristic impedance** of the line.

Suppose, next, that there is an observer at some point part way along the line (Fig. 7.4). Looking to the right, the observer sees a line exactly as if he had stood at the driven end. Thus the relationship between the current and the voltage remains as above.

Finally, suppose that the far end of the line has an impedance Z_0 connected across

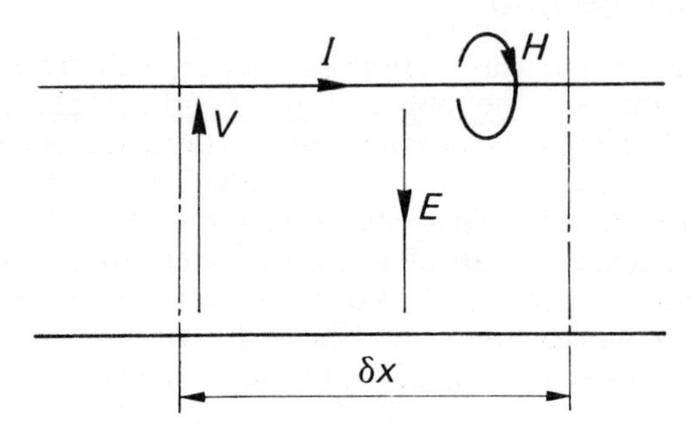

Fig. 7.2 Transmission line principle

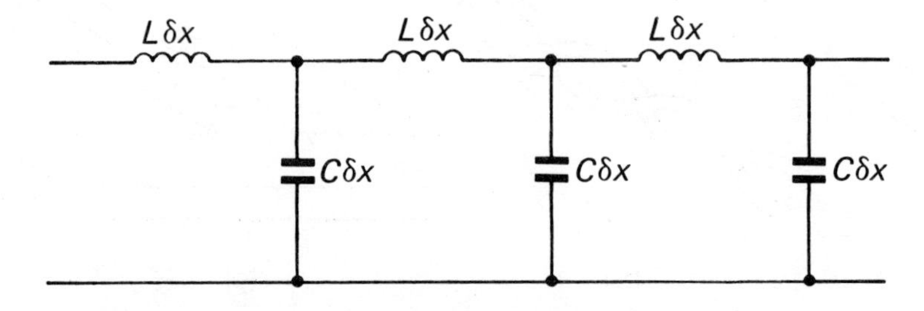

Fig. 7.3 Representation of a transmission line

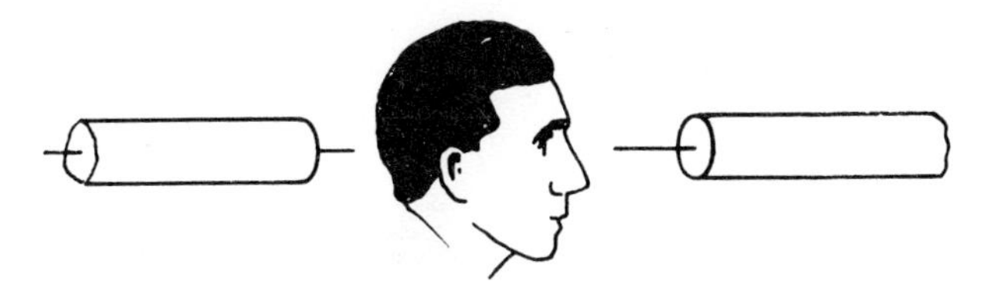

Fig. 7.4 Observer within a line

it. This looks like another piece of line. Hence this particular arrangement has the properties of an infinitely long line. The line is said to be **matched**.

To deduce the value of Z_0, it is most convenient to use the linear 'jω' technique. Consider one section of the line of Fig. 7.3. This feeds a further section, but that can be represented by the impedance Z_0 as shown in Fig. 7.5. Furthermore, an observer at the left of the section of line also sees an impedance of Z_0. Hence

$$\frac{V_i}{I} = Z_0$$

$$= j\omega L\delta x + \frac{1}{\dfrac{1}{Z_0} + j\omega C\delta x}$$

$$Z_0 - j\omega L\delta x = \frac{Z_0}{1 + j\omega C\delta x Z_0}$$

$$Z_0 - j\omega L\delta x + j\omega C\delta x Z_0^2 + \omega^2 LCZ_0\delta x^2 = Z_0$$

Dividing by $j\omega C\delta x$,

$$Z_0^2 - j\omega L\delta x Z_0 - \frac{L}{C} = 0$$

and as $\delta x \to 0$,

$$Z_0 \to \sqrt{\frac{L}{C}}$$

This is independent of frequency, and hence is a *pure resistance*.

Consider again Fig. 7.5. Remembering that the impedance seen from the input is Z_0

$$V_o = V_i \frac{1}{\dfrac{\dfrac{1}{Z_0} + j\omega C\delta x}{Z_0}}$$

$$\frac{V_o}{V_i} = \frac{1}{1 + j\omega C\delta x Z_0}$$

The phase of V_o relative to V_L is thus arctan $\omega C\delta x Z_0$. As δx becomes small, tan A tends to A, and hence

$$\theta \to \omega C\delta x\sqrt{\frac{L}{C}} = \omega\sqrt{LC}\delta x$$

For a sine wave, $\theta = \omega t$. Hence

$$t = \frac{\theta}{\omega} = \delta x\sqrt{LC}$$

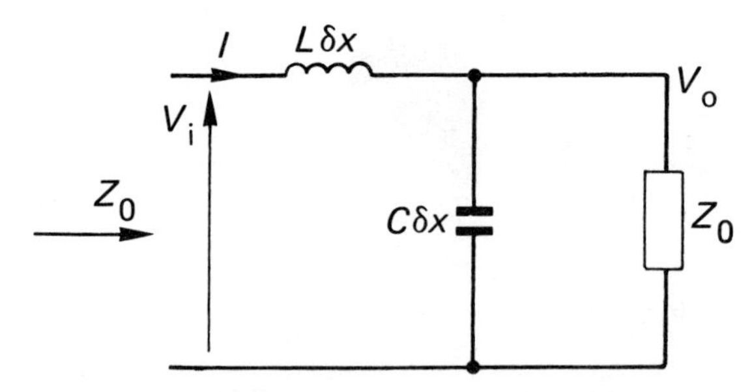

Fig. 7.5 Single section of a transmission line

This is a pure time delay, and again is independent of frequency. In particular, Fourier analysis shows that an edge can be regarded as the sum of an infinite series of sine waves. All of these are delayed by the same *time*, and hence the edge is undistorted. From the last formula it will be seen that

$$\text{delay per unit length of line} = \sqrt{LC}$$

As delay is the reciprocal of velocity,

$$\text{velocity} = \frac{1}{\sqrt{LC}}\,\text{ms}^{-1} = \frac{1}{\sqrt{\varepsilon\mu}} = \frac{1}{\sqrt{\varepsilon_0\varepsilon_r\mu_0\mu_r}}$$

For most materials, $\mu_r = 1$. However, most insulating materials used in transmission lines have values of ε_r of about 3 or 4. Thus the velocity of signals is around 50 to 60% of that in free space.

Termination of transmission lines

It has been pointed out earlier that a line terminated by an impedance of $\sqrt{L/C}$ looks like an infinite line. What happens with other terminations is one of the most difficult ideas in electronics. It is possible to demonstrate the effects mathematically, and the details will be found in texts on the subject; e.g. Mattick (see Appendix A). In this section a descriptive treatment will be given. The most convincing demonstration of the truth of the results is to observe the effects on a real line. This can be done most easily with a suitable square wave generator driving a piece of coaxial or twisted pair cable, and an oscilloscope. If the cable is of the order of 30 metres in length, the effects are observable on a 10 MHz oscilloscope.

Open circuit line

Figure 7.6 shows an open circuit line driven by a square wave voltage generator via an impedance of R_0 (the reason for this last will appear shortly). As the impedance of the line is also R_0, the current into the line will be given by

$$I = \frac{V}{2R_0}$$

and the voltage at the input to the line is $V/2$. Call this the **forward wave**. Waveforms are shown in Fig. 7.7.

The signal travels down the line, and eventually reaches the termination. As this is an open circuit, the current must be zero. In order to achieve a zero current, imagine a second current wave travelling in the opposite direction to the forward wave, and arriving at the termination at the same time. The forward wave is imagined to con-

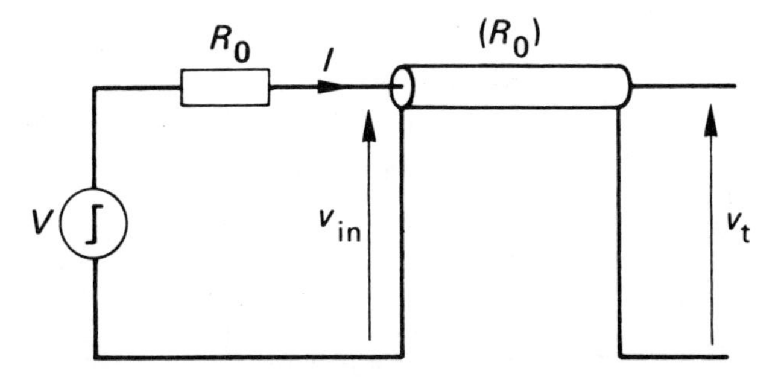

Fig. 7.6 Open circuit line

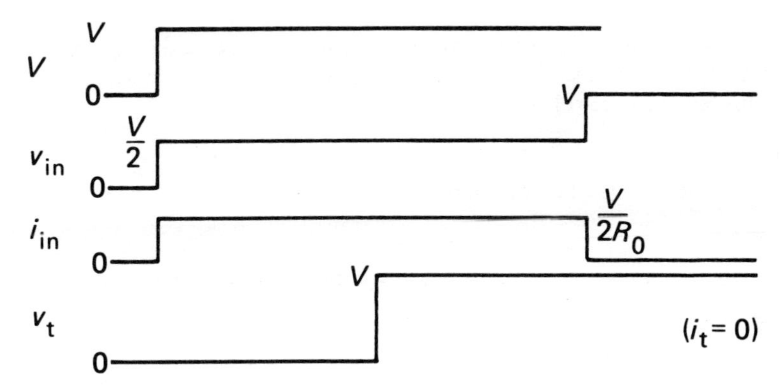

Fig. 7.7 Waveforms in an open circuit line

tinue to infinity, and the new wave travels back towards the source. There is thus no current actually flowing in the line. However, the new current does flow in the impedance of the line, and generates a voltage of IR_0, the same as that of the forward wave. Hence the voltage at the termination is *double* that arriving. The **backward wave** is directly comparable with the reflected ray of light from a mirror, and is termed a **reflection**. If the ray of light strikes the mirror normally, the intensity of the total light will be doubled.

To be complete, the backward wave now continues along the line, and eventually arrives at the source end. This is 'terminated' by R_0, and so appears like the infinite line. There is no further 'reflection'. It will now be appreciated why the source resistance of the generator was chosen to be R_0. The idea of the second wave can be difficult to understand.

The final condition can be confirmed by the following procedure. Once the transmission line effects have died away, the line itself can be treated as a connection of zero resistance. The voltage on this wire is V, and there is no current flowing, since an open circuited line has an infinite load impedance. In the previous description the forward and backward waves had a magnitude of $V/2$, giving a total of V. The current was proposed in such a way as to give a resultant value of zero.

The general case of termination

Figure 7.8 shows a line driven by a voltage source of resistance R_s, and terminated by a resistance of R_t. The input voltage and current are v_{in} and i_{in}, and those at the

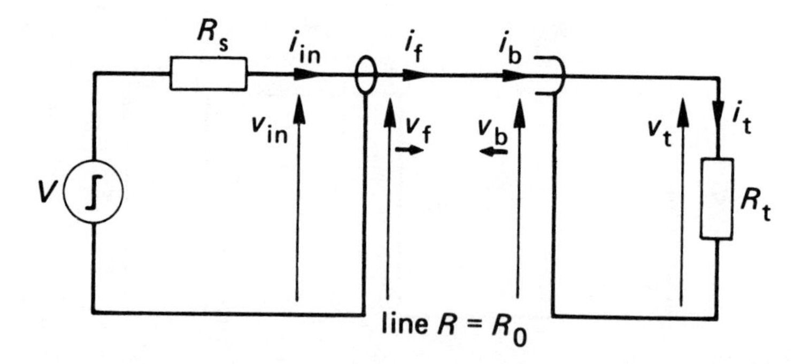

Fig. 7.8 Line terminated by an arbitrary resistance

termination are v_t and i_t. The first forward wave is characterised by v_f and i_f, and the backward wave by v_b and i_b. The relevant directions of flow are indicated by arrows in Fig. 7.8. This leads to the relations

$$v_f = i_f R_0$$

$$v_b = -i_b R_0 \qquad \text{(notice the sign!)}$$

$$v_t = i_t R_t$$

Consider the situation at the termination when the forward wave arrives.

$$v_t = v_f + v_b$$

$$i_t = \frac{v_f}{R_0} - \frac{v_b}{R_0}$$

and $\quad i_t = \dfrac{v_t}{R_t}$

From these equations,

$$\frac{v_f}{R_0} - \frac{v_b}{R_0} = \frac{v_f + v_b}{R_t}$$

$$v_b\left(\frac{1}{R_0} + \frac{1}{R_t}\right) = v_f\left(\frac{1}{R_0} - \frac{1}{R_t}\right)$$

$$\frac{v_b}{v_f} = \frac{R_t - R_0}{R_t + R_0}$$

$$= \Gamma$$

The factor Γ (gamma) is called the **voltage reflection coefficient**. Now

$$v_t = i_t R_t$$

$$= (i_f + i_b)R_t$$

$$= \frac{(v_f + v_b)R_0}{R_0}$$

$$= (i_f - i_b)R_0$$

$$(i_f + i_b)R_t = (i_f - i_b)R_0$$

$$\frac{i_b}{i_f} = \frac{R_0 - R_t}{R_0 + R_t} = -\Gamma$$

Thus the **current reflection coefficient** is the negative of the voltage reflection coefficient.

Considering the case of a matched line, $\Gamma = 0$, and hence $v_b = 0$ and $i_b = 0$ as might be expected. For the open circuit line, R_t is infinite. Hence $\Gamma = 1$, $v_b = v_f$, and $i_b = -i_f$. Thus $v_t = 2v_f$ and $i_t = 0$, as found earlier.

When the backward wave returns to the source, a further reflection takes place. To find the new forward wave the same formulae are used, with R_s as the 'terminating' resistance.

To find the value of a waveform at any time, it is only necessary to add all the components. However, in calculating the components it is essential to remember that it is only the incident wave that is reflected, and not the 'total so far'. The final values can always be calculated by treating the line as a connection of zero resistance. This is a useful check on other calculations, and also to indicate when the 'total so far' is sufficiently close to the final value.

Short circuit line

To fix ideas, a number of examples will be taken. The first of these is the short circuit line shown in Fig. 7.9. The source is again *matched* by the resistance R_0. Thus the forward wave is the same as for the open circuit line. In this case $\Gamma = -1$, and hence $v_b = -v_f$, giving a value of zero for v_t, which is what would be expected of a short circuit. However, $i_t = 2i_f$.

The backward wave now returns to the source, which is matched. The final value of current is V/R_0, and v_{in} is zero, again as expected from treating the line itself as a zero resistance connection. Figure 7.10 shows the waveforms.

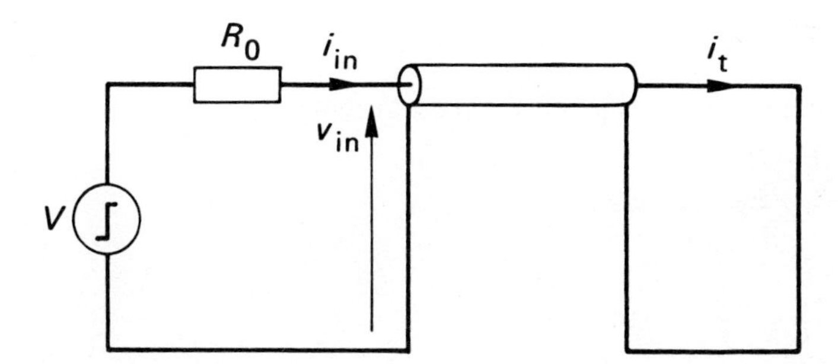

Fig. 7.9 Short circuit line

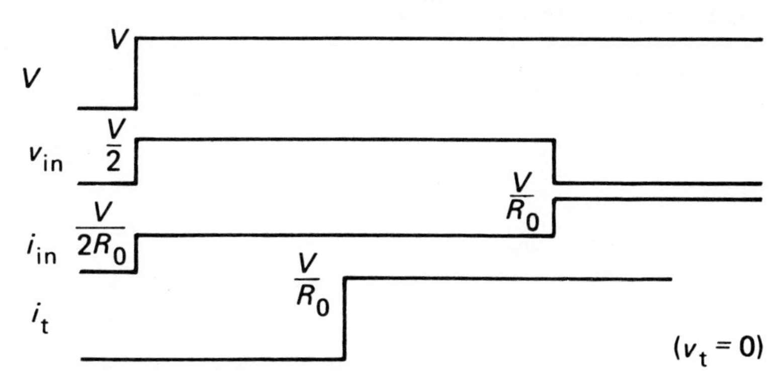

Fig. 7.10 Waveforms for the short circuit line

Illustration of the general case

To illustrate the general case, the value of R_t is chosen as $2R_0$, and R_s is chosen to be $R_0/2$. There is nothing magical about these numbers. They are chosen as one larger and one smaller than R_0, and sufficiently different from R_0 to demonstrate the effects. Figure 7.11 shows the circuit.

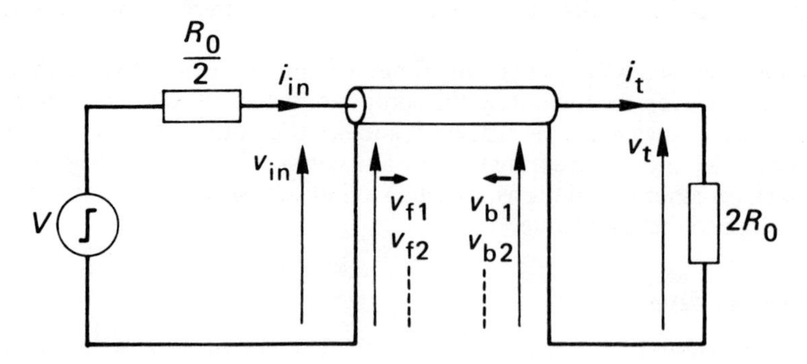

Fig. 7.11 'General' termination for a line

Starting with the input waveform,

$$i_{in1} = \frac{V}{R_s + R_0} = \frac{2V}{3R_0} = i_{f1}$$

$$v_{in1} = \frac{2V}{3} = v_{f1}$$

At the termination, the voltage reflection coefficient is

$$\Gamma_t = \frac{R_t - R_0}{R_t + R_0} = \frac{1}{3}$$

and hence

$$v_{b1} = \frac{v_{f1}}{3} = \frac{2V}{9}$$

$$i_{b1} = -\Gamma i_{f1} = -\frac{2V}{9R_0}$$

Thus the conditions in R_t at time $l\sqrt{LC}$ are

$$v_t = v_{f1} + v_{b1} = V\left(\frac{2}{3} + \frac{2}{9}\right)$$
$$= \frac{8V}{9}$$

$$i_t = i_{f1} + i_{b1} = \frac{2V}{3R_0} - \frac{2V}{9R_0}$$
$$= \frac{4V}{9R_0}$$

and thus $v_t/i_t = 2R_0$ as expected. After a further time of $l\sqrt{LC}$, the backward wave reaches the source, and is reflected again. Here we have

$$\Gamma_s = \frac{R_s - R_0}{R_s + R_0} = -\frac{1}{3}$$

$$v_{f2} = -\frac{1 v_{b1}}{3} = -\frac{2V}{27}$$

Notice that this is $-1/3$ of v_{b1}, not $-1/3$ of v_t.

$$v_{in2} = v_{f1} + v_{b1} + v_{f2}$$

$$= V\left(\frac{2}{3} + \frac{2}{9} - \frac{2}{27}\right)$$

$$= \frac{22V}{27}$$

$$i_{f2} = +\frac{1 i_{b1}}{3}$$

$$= -\frac{2V}{27R_0}$$

$$i_{in2} = \frac{V}{R_0}\left(\frac{2}{3} - \frac{2}{9} - \frac{2}{27}\right)$$

$$= \frac{10V}{27R_0}$$

This procedure is now continued for as long as is necessary. However, the final values can be obtained by treating the transmission line as a connection of zero resistance. Under these conditions R_t and R_s form a simple potential divider. Hence

$$V_{in} = V_t = \frac{4V}{5} = \frac{21.6V}{27}$$

$$I_{in} = I_t = \frac{2V}{5R_0} = \frac{10.8V}{27R_0}$$

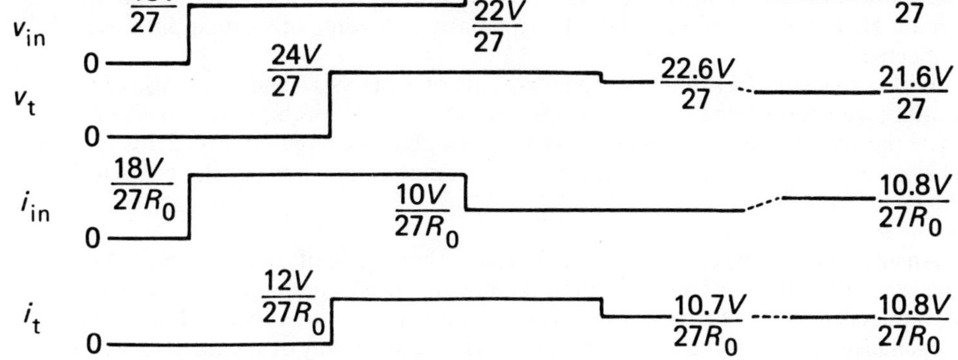

Fig. 7.12 Waveforms for the circuit of Fig. 7.11

Comparing the final values of V_{in} and I_{in} with the totals so far, v_{in2} and i_{in2}, it will be seen that the values are already quite close. This is also an indication that the result so far is likely to be correct. Figure 7.12 shows the waveforms.

Equivalent circuit of a line and termination

Figure 7.13 shows an equivalent circuit of a transmission line and its termination. A signal v_f is traversing the line, and when it reaches the termination it is reflected. The total terminal voltage is given by

$$v_t = v_f\left(1 + \frac{Z_t - Z_0}{Z_t + Z_0}\right)$$

$$= \frac{v_f 2Z_t}{Z_t + Z_0}$$

from Fig. 7.13, the value of v_t is

$$v_t = \frac{Z_t}{Z_t + Z_0} v$$

As these must be equal, $v = 2v_f$. The reader can confirm that this equivalent circuit gives the same results as previous methods.

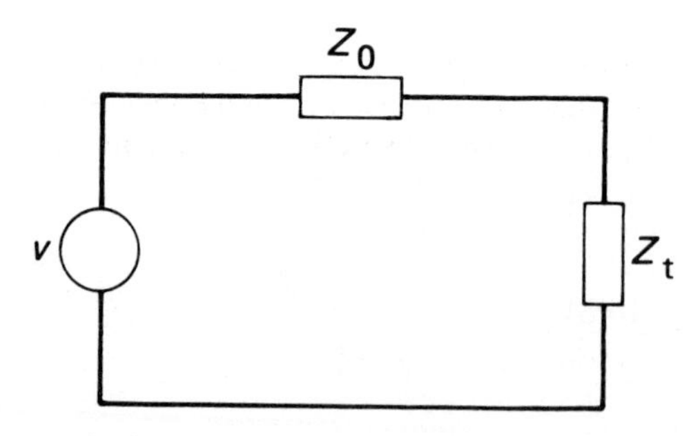

Fig. 7.13 Equivalent circuit of a line and termination

Capacitive termination

In real circuits the input impedance of a gate is several kilohms, and the value of R_0 is generally of the order of 50 to 150 Ω. Thus the gate input can be treated as an open circuit. However, in high speed work the capacitive loading of a gate input cannot be neglected.

With a capacitive termination, the mathematical analysis requires the use of the Laplace transform value of Z_t, which is beyond the scope of this text. However, by using the circuit of Fig. 7.13 with Z_t as a capacitance, the circuit of Fig. 2.1 is formed. This gives an output which is an exponential change with a time constant CR_0 to $2v_f$ as shown in Fig. 7.14.

An alternative view of this is to realise that a sharp edge of a waveform can be regarded as the result of a very high frequency sine wave component in the signal. Now the impedance of a capacitance is $1/\omega C$, and hence at this very high frequency the capacitance appears to be a short circuit. This is another way of saying 'the voltage

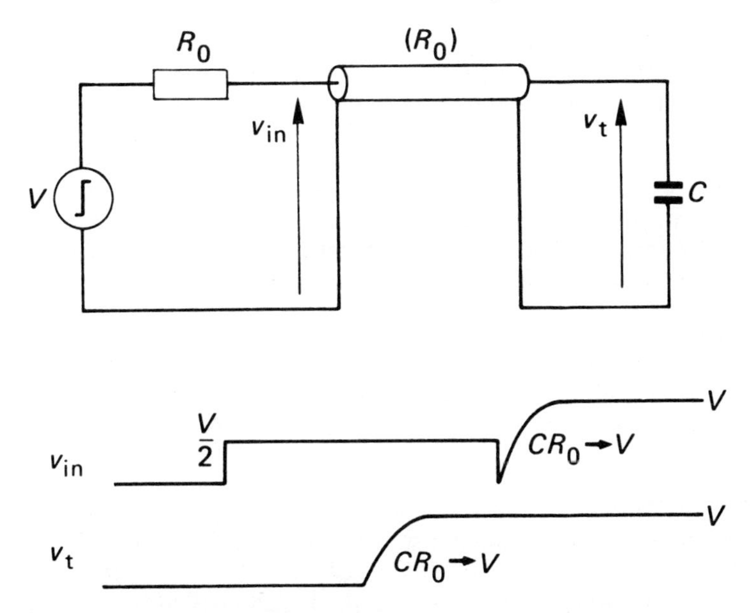

Fig. 7.14 Capacitive termination ($R_s = R_0$)

across a capacitance cannot be changed instantaneously'. Thus, if a transmission line is terminated by a capacitance, it can be treated *initially* as a short circuit termination.

After a 'long time', when all transient changes have died away, the capacitance appears as an open circuit. Thus the voltage across it is V.

Figure 7.14 shows the waveforms when the line is matched at the source. v_{in} is initially $V/2$. This is reflected at the termination, giving $v_{b1} = -V/2$. There is no reflection at the source, since that end of the line is matched. The final value of both v_{in} and v_t is V, since the capacitance is an open circuit to steady voltages.

A similar approach can be taken for an inductive termination. In this case the inductance appears initially as an open circuit, and after a 'long time' as a short circuit (or a short circuit in series with a resistance).

When is a connection a transmission line?

The consideration of the 'general' case shows that even for quite severe mismatch conditions, transmission once down the line and back is sufficient for conditions to reach 90% of the final value. Hence as a rough and ready rule, a wire need not be treated as a line if the time for a signal to travel down the line and back is less than the rise time of the signal. This occurs if the line is short. It should be borne in mind that for a circuit with a 1 ns rise time, wires must be less than about 7 mm (including those within IC packages) to be regarded as 'short'. For signals with 10 ns rise times, distances of up to 0.75 m are acceptable as being short.

Driving lines with real logic circuits

Transmission lines usually have characteristic impedances in the range 50 to 150 Ω. This is a low value in electronic terms, and hence driving circuits must be capable of delivering substantial current. To avoid reflection problems, the line must also be matched under all logic conditions. Thus the source and terminating impedances must either be sensibly constant, or they must be so high in relation to Z_0 that they can be treated as open circuits.

Considering the common circuit families, TTL does not meet the requirements. The output impedance of a normal gate is around 10 Ω in the low logic state, and over 100 Ω in the high state. Thus these circuits cannot be matched to a line for both states at the same time. For driving lines there are several special buffers with low output impedances. Even here, however, care must be taken, as many of these buffers can only drive a line of characteristic impedance of greater than 133 Ω. This value is high in transmission line terms. It is also possible to use **open collector** gates with a matching resistor to ground, or to use **three state** (tristate) devices, where the output impedance can be made to go very high. Matching is then achieved by means of external resistors.

With the high speed ECL circuits the problems are easier. The output circuit is an emitter follower that should never turn off, and hence the output impedance is always low—less than 10 Ω. This is much smaller than Z_0, and hence matching is quite easy using external resistors.

Series matching

There are two methods of matching transmission lines; series and parallel. Figure 7.15 shows a series matched line. A low output impedance gate, such as an ECL device, drives the line via a matching resistor. The load is high impedance, and can be regarded as an open circuit or as a pure capacitor, as appropriate. The waveforms for this circuit are those of Fig. 7.7 or Fig. 7.14.

With this type of matching, all loads must be placed at the termination. The separation must obey the rule that the time for a signal to travel between the two driven gates furthest apart must be less than half the rise time. If this is not the case, then there are two problems.

Consider the extreme case of a load placed near the source. Firstly, the input signal (v_{in} of Figs 7.7 and 7.14) only goes to $V/2$ initially. For ECL this means that the output may also go to $V/2$. In the presence of noise, or for TTL, the output may oscillate. Both circumstances are highly undesirable. Secondly, even without these effects, it would still take twice the time for the input to change by V (see Figs 7.7 and 7.14). This may be acceptable in a non-critical place, but not in general.

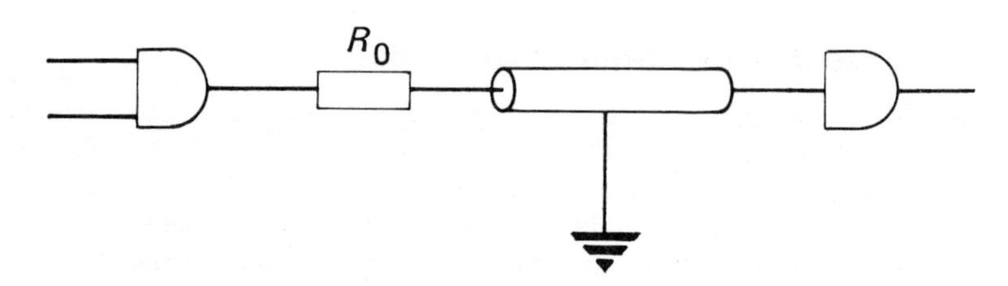

Fig. 7.15 Series matched line

Parallel matching

Figure 7.16 shows a parallel matched line. In this case the line should be driven by a gate with a low output impedance, or by a current source (very high output impedance). A full amplitude signal enters the line. A resistor equal to R_0 is connected at the far end, and appears like an infinite line. The resistor must be returned to a voltage level appropriate to the circuit of the gates. For example, with ECL gates, the resistor will normally be connected to a −2 V supply (output voltage levels are −0.85 V and −1.7 V).

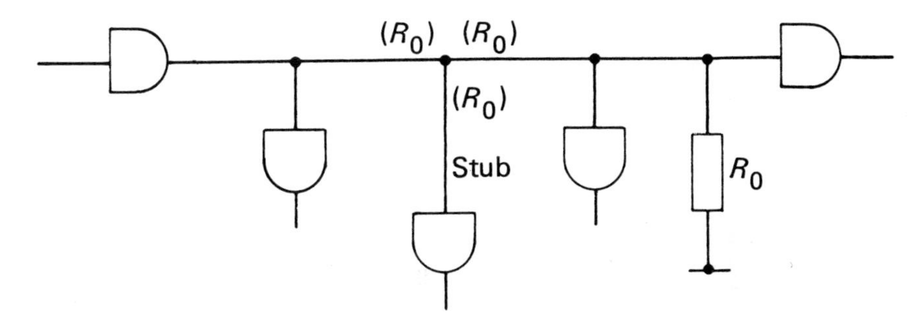

Fig. 7.16 Parallel matched line

In this case the loading gates do not need to be at the end of the line. However, two precautions need to be observed.

Firstly, the 'stubs' from the line to the gate must be short (see Fig. 7.16). If they are not, then the stub becomes a second line. At the point where the line divides, the signal sees an impedance of $R_0/2$ (two lines of impedance R_0 in parallel), and hence there is a reflection.

Secondly, the loads must be evenly distributed. The capacitance of the gate input adds to that of the line. If the loads are evenly distributed, this capacitance can be added to that of the line, thus reducing the value of $R_0(\sqrt{L/C})$, and increasing the line delay ($\sqrt{LC}$). If the loads are unevenly distributed, the effective impedance of the line varies, with consequent effects on the waveforms. Of course, all loads may be placed together near the termination.

Comparison of series and parallel matching

Speed

It was shown earlier that a series matched line terminated by a capacitive load has a time constant of CR_0 associated with it (Fig. 7.14). With a parallel matched line, the resistor R_0 is across the capacitance. Hence the time constant is $CR_0/2$ (see Figs 2.9 and 7.14). Thus the parallel matched system seems to be faster (but see the paragraph entitled *Number of lines* below).

Power dissipation

If the transmission line is properly matched, the maximum amount of power passes to the load. The amount of power dissipated in other components varies with the matching arrangements. The initial current into a series matched line is $V/2R_0$, since the matching resistor adds to the impedance of the line. However, in order to cause the opposite change, significant power is needed. Consider Fig. 7.17. When the transistor switches on, the impedance is low, and the line is matched. Power dissipation (for a saturated transistor) is V_s^2/R. When the transistor switched off, voltage V must be switched immediately. The current into the line is $V/2R_0$, and at least this must flow in R. Thus

$$\frac{V_s - V}{R} > \frac{V}{2R_0}$$

$$R < 2R_0\left(\frac{V_s}{V} - 1\right) \tag{7.1}$$

To get a reasonable value for R, V_s/V must be at least 2.

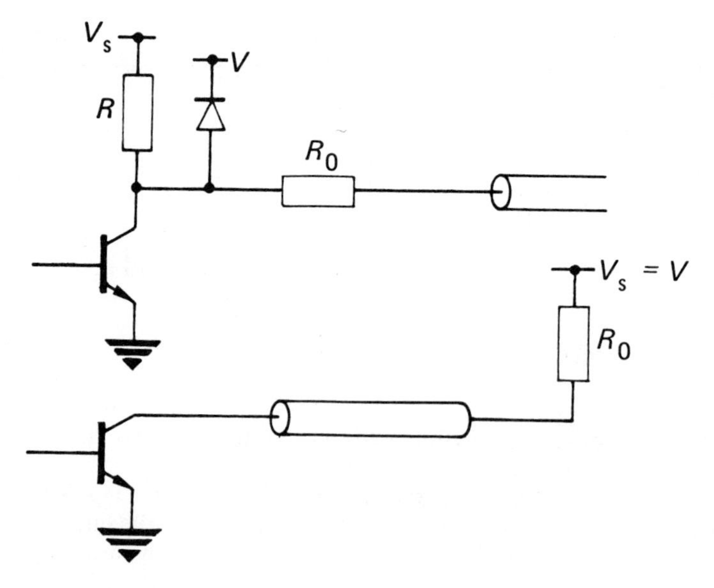

Fig. 7.17 Line driving circuit

Suppose that V_s/V is 3, and that the maximum voltage into the line is V. Then the power dissipation in this state is

$$\frac{(V_s - V)^2}{R} = \frac{4V_s^2}{9R}$$

Thus the total dissipation, assuming half the time in each state, is

$$\text{Power} = \frac{1}{2}\frac{V_s^2}{R}\left(\frac{4}{9}+1\right) = \frac{13}{18}\frac{V_s^2}{R} = \frac{13}{2}\frac{V^2}{R}$$

For the parallel matched case, V_s need be no larger than V, and R is, in fact, R_0 but placed at the termination. The current into the line will be V/R_0, twice that for the series matched case. The power dissipation is V^2/R_0 in one state, and very small in the other. Since V_s is not equal to V (e.g. for ECL $V_s = -2$ V, $V = -1.7$ V), the power dissipated will be rather greater than $V^2/2R_0$. At the worst, it will be $3V^2/4R_0$, say.

Now R must be less than $4R_0$ in the series matched case (see equation 7.1). Suppose $R = 3R_0$. Then

$$\text{power in series matched case} = \frac{13V^2}{6R_0}$$

$$\text{power in parallel matched case} = \frac{3V^2}{4R_0} = \frac{4.5V^2}{6R_0}$$

Thus the power in the parallel matched case is much less. If V_s is 2 V, then the power becomes comparable, but R becomes quite small.

Number of lines

Since the parallel matched line requires twice the current drive of the series matched case, it is possible to drive twice as many lines in the latter case. This has two effects for the series matching.

Firstly, the value of R is reduced, and hence the power dissipation in the series matched case will increase.

Secondly, the loads can be split between twice as many lines. Hence the capacitance *per line* is halved, and the time constants for series and parallel systems with the same total number of loads will be the same.

Choice between series and parallel matching

The choice between series and parallel matching is not simple. The ability to distribute loads in the parallel case can be very useful, but the need for short stubs may result in a much longer path from source to the most distant load. It is also important to distribute the loads correctly. In a recent practical case, the author distributed a set of loads over the second half only of a parallel matched line, for simplicity of layout. The result was a rise time of the signals 2 to 3 times slower than needed at the load nearest the source. On the other hand, the need to place all the loads close together in the series matched case is restrictive. With circuits of less than 1 nanosecond, 'close' means less than the length of one IC.

Driving a busbar

The different units of many modern computers are interconnected by means of a busbar or **bus** structure. The distance between units is usually large enough for the bus lines to have to be treated as transmission lines. Figure 7.18 shows one line of such a system. The stubs to each unit are assumed to be 'short'.

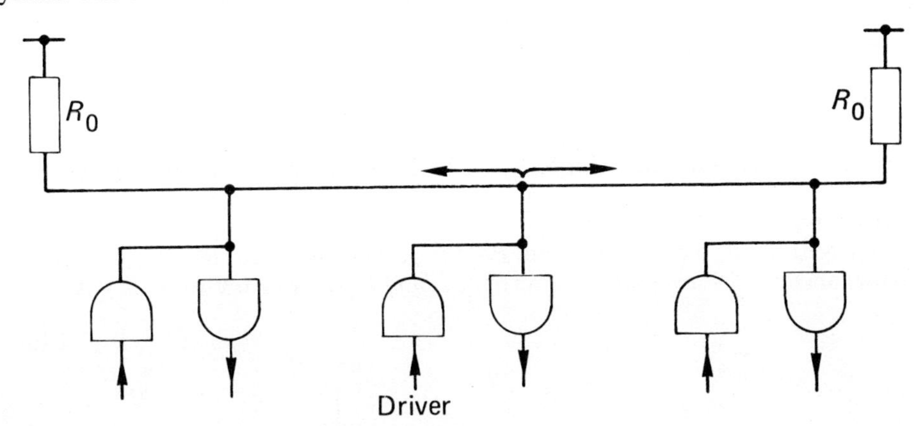

Fig. 7.18 One wire of a bus system

Suppose a unit near the middle of the line acts as the driver. The signal will travel *both ways* down the wire. If there is to be no reflection, there must be a matching resistor at *both* ends of the line.

To calculate the value of the matching resistor (or the line dimensions for a required R_0), the value of the capacitive load of each unit must be added in. Suppose an impedance of 100 Ω is required, the capacitance per load is 4 pF, and there are 8 loads per metre. Then the added capacitance per metre is 32 pF. One might expect to achieve an inductance of 0.7 μH m^{-1}. Thus

$$\sqrt{\frac{L}{C}} = 100 = \sqrt{\frac{0.7 \times 10^{-6}}{C}}$$

$$C = 0.7 \times 10^{-6} \times 10^{-4}$$

$$= 70\,\text{pF}$$

But the load is 32 pF m^{-1}. Hence the line must have an inherent capacitance of 38 pF m^{-1}.

It is unlikely that these figures can be achieved easily, and the design may require several iterations to produce a system that can be manufactured.

Real matching

TTL circuits

With TTL drivers, the circuits must either be tristate devices, or open collector devices. In either case the matching resistors must set the high logic level. A 100 Ω resistor connected directly to +5 V would result in 50 mA flowing in T, which is more than it is capable of carrying. To reduce this, the high level is set to 3 V. Suppose that an impedance of 100 Ω is required.* Figure 7.19 illustrates the resistors. The line must be at 3 V when not being driven, and the resistors *in parallel* must have a value of 100 Ω. Thus

$$\frac{R_1 5}{R_1 + R_2} = 3$$

$$2R_1 = 3R_2$$

and
$$\frac{R_1 R_2}{R_1 + R_2} = 100$$

$$\frac{3R_2^2}{2} = 100\frac{5R_2}{2}$$

$$R_2 = \frac{500}{3} \simeq 180\,\Omega$$

to the nearest preferred value. Hence $R_1 = 270\,\Omega$, which is also a preferred value.

ECL circuits

ECL may be series or parallel matched. For the parallel matched case a process similar to that above will show that resistors of 82 Ω and 120 Ω from 0 V to −5.2 V is

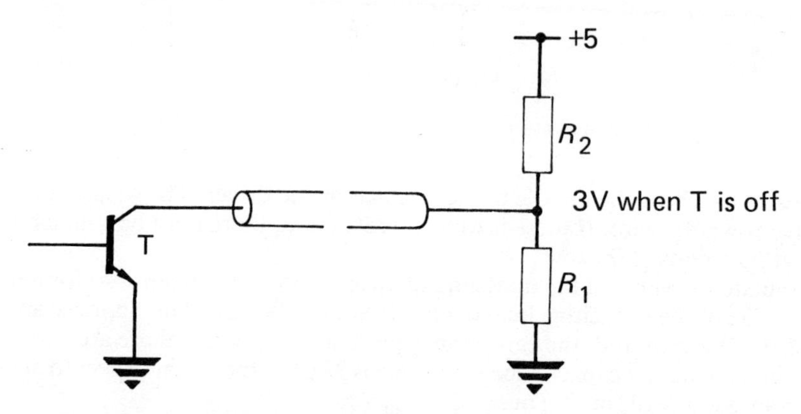

Fig. 7.19 Matching resistors for a bus (parallel match)

* Several of the TTL circuits intended for bus driving require lines of 133 Ω or *more*. This is a high value in transmission line terms. Further, the use of matching at both ends of the line as indicated above is not possible. Great care must be exercised in choosing circuits.

equivalent to 50 Ω to −2 V. 120 Ω and 180 Ω is equivalent to 75 Ω, and 180 Ω and 270 Ω leads to 100 Ω matching.

For series matching a pull down resistor is required. This can be calculated using the technique in the section dealing with power dissipation. Thus

$$\frac{5.2-1.7}{R} > \frac{0.9n}{2R_0}$$

when n is the number of lines, and the voltage swing is 0.9 V. Hence

$$R < \frac{3.5 \times 2R_0}{0.9n} = \frac{8R_0}{n}$$

To drive two 75 Ω lines, R must be 270 Ω or less (nearest preferred value below 300 Ω).

ECL is designed to drive at least one 50 Ω transmission line. Many circuits can drive two, or one 25 Ω line.

Use of a short circuited transmission line for pulse generation

The short circuited line can be used as a generator of pulses, especially when they need to be very short, and with fast edges. Figure 7.20 shows the circuit and the waveforms.

A voltage generator giving an edge of amplitude E volts drives a line of characteristic impedance R_0 via a matching resistance R_0, the line is terminated by a short circuit. The output of the circuit is at A, the input to the line. The initial current into the line is $E/2R_0$, and the voltage at A is $E/2$ volts. The reflection at the termination produces a backward voltage wave of $-E/2$, and a current wave of $E/2R_0$. When this reaches A the voltage reduces to zero, and the final current is E/R_0. There is no further reflection, since the source is matched. The pulse is $2l\sqrt{LC}$ seconds long. The pulse

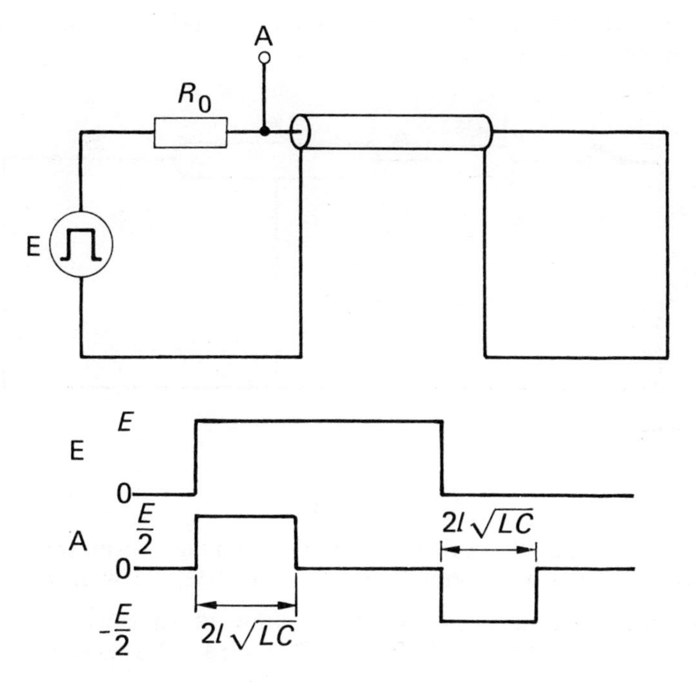

Fig. 7.20 Pulse generation using a short circuited transmission line

amplitude is $E/2$ volts. When the other edge of the input occurs, a second pulse is generated, with amplitude $-E/2$ volts and of the same length.

Two points should be noted. Firstly, the pulses are only half the amplitude of the input. They may not be able to drive the load if this is a normal gate. Thus, some care in use is required.

Secondly, one cycle of the input produces *two* pulses of opposite polarity. This may not matter, but if the input of a gate is taken negative with respect to the lowest voltage in the circuit, the IC may fail to work correctly.

In certain circumstances the double pulse is an advantage. Regardless of the number of input cycles or their mark–space ratio, the area under the waveform integrates to zero. Thus, the total charge driven into the output is zero. This is important when driving very long lines—for example telephone lines, especially if they are submarine.

An interesting special case of this pulse generator is when the input pulse is also $2l\sqrt{LC}$ seconds in length. In this case the leading edge of the negative going pulse is coincident with the trailing edge of the positive going pulse.

Conclusions

This chapter demonstrates some of the effects of transmission lines in digital systems, without covering the detailed mathematics involved. There are useful graphical techniques to assist in determining the effects of mismatch. The principles described here should be sufficient to enable good designs to be achieved in the vast majority of cases, provided that reasonable care is exercised. The reader is warned against being complacent and regarding lines as short circuits. With slow or uncritical circuits this may be possible, but each case should be properly assessed. As circuit speeds get higher, fewer and fewer wires can be treated as short circuits rather than as transmission lines.

Questions

1 A transmission line has $L = 0.82\,\mu\text{H}\,\text{m}^{-1}$ and $C = 41\,\text{pF}\,\text{m}^{-1}$. What is the relative permittivity of the dielectric?

2 

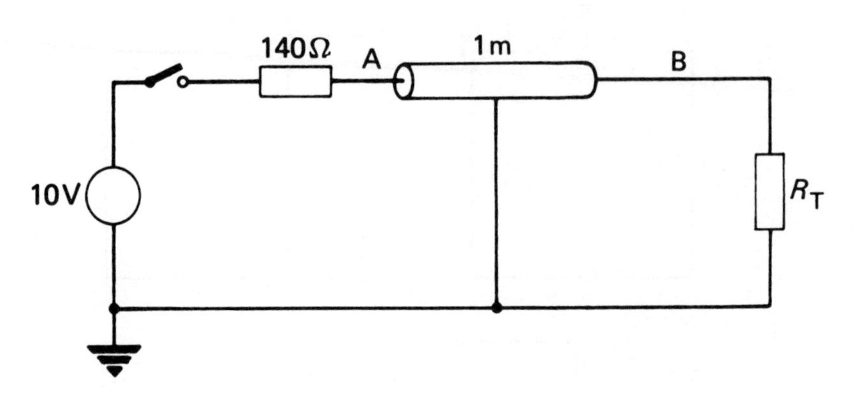

Fig. 7.21

The switch in Fig. 7.21 closes at time $t = 0$. Draw voltage waveforms at A and B if the transmission line is as in question 1, and $R_T =$ (a) 70 Ω; (b) 140 Ω; (c) 280 Ω.

3 With the same arrangement as in question 2, draw the waveforms at A and B if R_T is replaced by $Z_T =$ (a) 10 pF; (b) 10 pF in parallel with a 140 Ω resistor.

4 In the circuit of question 2, R_T is 70 Ω, input is 700 mV, and the switch is closed at

$t = 0$. C is halfway along the line which, in this case, is 10 ft long. The relative permittivity is 2.25, and the velocity of light in vacuo is 1 ft ns^{-1}. Draw waveforms at A, B and C.

5 A bus transmission line system is 5 m long, and is designed to allow communication between up to 25 independent units. Each unit has a transmitter/receiver circuit which can be plugged into the bus at any point. When connected, it loads the line with 5 pF.

If the unloaded lines have a propagation delay of 5 ns m^{-1}, and an impedance of 100 Ω, what is the effective bus delay and impedance when all 25 units are plugged in at regular intervals along the length of the bus?

6 Using gates and a delay line, design a circuit to generate a pulse of 10 ns width, and draw all relevant waveforms. What is the recovery time and maximum repetition rate?

7 Under what circumstances is it necessary to regard the connection between logic circuits as transmission lines?

What is the delay time of a signal travelling along a printed circuit board track (approximately)?

Explain how you may determine the input and output waveforms of a transmission line driven by a source of resistance R_s and terminated by a resistance R_T.

A transmission line of characteristic impedance 100 Ω is driven by a 1 V source of resistance 67 Ω and terminated by a resistance of 150 Ω. Determine the waveforms at the input and output of the line. U of M (May 1981)

8 Describe how the input and output waveforms of a transmission line may be estimated from general circuit considerations, without recourse to detailed analysis. Use your method to determine waveforms at input and output of a line in response to a step in the source voltage for the following cases:

(a) a line matched at the source, and terminated by a short circuit;
(b) a line matched at the source, and terminated by 50 kΩ in parallel with a capacitance;
(c) a line driven from a low impedance source and terminated by the line characteristic impedance in parallel with a capacitance.

Discuss the circuit and transmission line problems associated with the design of a bus interconnecting the various units within a computer system.

U of M (Sept 1981)

9 What are the two methods by which a pair of gates may be connected via a transmission line in order to minimise the operation time of the circuitry?

A transmission line 1 m long has a characteristic impedance of 75 Ω, and signals take 5.3 ns to travel from one end to the other. It is loaded by four gates each with an input impedance consisting of 50 kΩ in parallel with 4 pF, all connected close together. The line is driven by a gate whose output impedance is 7 Ω.

Sketch the circuit for both the methods of connection, giving values of additional components, and draw waveforms at source and termination resulting from a step change of voltage at the source. Describe how the waveform is derived.

How can the rise time at the termination of a series matched line be improved relative to the parallel matched line using the same driver, and the same total load?

What would be the consequence of spacing the loads equally along the line in the two cases? Give values for any components which need to be modified.

U of M (May 1980)

10 With the aid of waveform diagrams, explain what happens when a logic gate with an output impedance of 75 Ω drives a 'long' track on a printed circuit board to a second gate with a high input impedance (resistive).

If the logic gate has an output signal rise time of 2 ns, derive a value for the maximum length of wire that can sensibly be driven without a matching resistor. If the logic gate input impedance has, in fact, a significant capacitance, what would be the waveforms at the two ends of the wire? Explain how these waveforms might be derived from the characteristics of the capacitor.

How would your waveforms be affected by the type of matching used (series or parallel)? U of M (Sept 1982)

11 A signal V volts is being transmitted down a transmission line of characteristic impedance R_0. The line is terminated by an impedance Z_T. Derive an equivalent circuit representing the conditions at the termination when the signal arrives there.

If Z_T is a capacitance in parallel with a 55 kΩ resistor, determine the rise time of the signal at the termination, assuming that the rise time of the original forward wave is small in comparison.

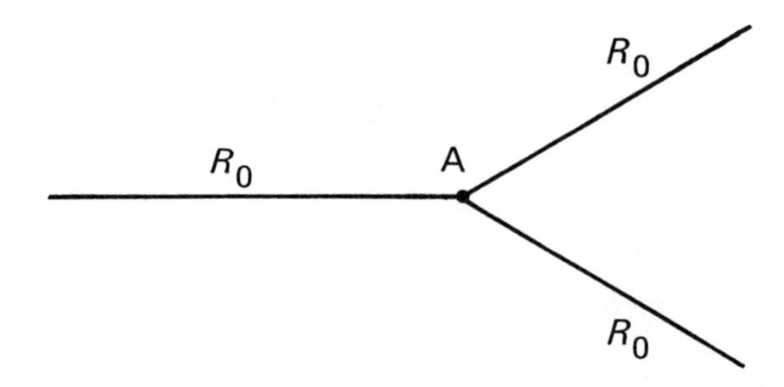

Fig. 7.22

A signal V travels along a line of impedance R_0, which branches at a point A as shown in Fig. 7.22. Using previous results, determine the waveform at A.

What voltage travels down each of the two branches?

Confirm your results making use of the concept of a reflection at a point A, and the reflection coefficient. U of M (June 1983)

8 Ramp generators

Introduction

A **ramp** is a waveform for which the voltage (or current) changes linearly with time. The instantaneous value of the voltage is thus a measure of time. Use is made of this fact in a number of places, notably as the time base of an oscilloscope (or the sweep generator which scans the spot across the screen of a television). This chapter discusses some of the principles governing the design and use of ramp generator circuits.

Definitions

Figure 8.1(a) shows a real ramp waveform. The non-linearity is deliberately exaggerated for clarity. In practice, the curvature would not be noticeable to the unaided eye. The figure gives a diagrammatic definition of a number of terms. The minimum reset time is also shown.

The linearity of a ramp is an important parameter. It is expressed as the deviation from linearity, given by

$$\text{deviation from linearity} = \frac{\text{initial slope} - \text{final slope}}{\text{initial slope}}$$

It is most frequently expressed as a percentage.

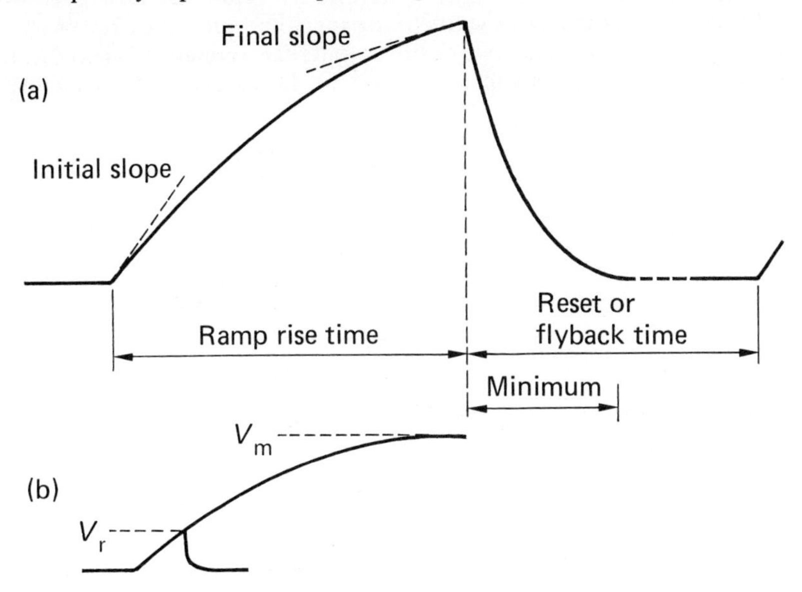

Fig. 8.1 Definitions relating to ramp waveforms: (a) real ramp—curvature exaggerated, (b) deviation from linearity

In most cases a ramp is, in fact, a part of an exponential waveform as shown in Fig. 8.1(b). The actual ramp is only a small part of the maximum possible change, and hence is almost linear. From Fig. 8.1(b)

$$v = V_{\mathrm{m}}(1 - \mathrm{e}^{-t/T}) \tag{8.1}$$

$$\frac{\mathrm{d}v}{\mathrm{d}t} = \frac{V_{\mathrm{m}}}{T}\,\mathrm{e}^{-t/T}$$

For the initial and final slope, substitute $t = 0$ and $t = t_{\mathrm{r}}$ in the equation.

At $t = 0$,

$$\frac{\mathrm{d}v}{\mathrm{d}t} = \frac{V_{\mathrm{m}}}{T}$$

At $t = t_{\mathrm{r}}$, $v = v_{\mathrm{r}}$, and

$$\frac{\mathrm{d}v}{\mathrm{d}t} = \frac{V_{\mathrm{m}}}{T}\,\mathrm{e}^{-t_{\mathrm{r}}/T}$$

$$\text{deviation from linearity} = \frac{V_{\mathrm{m}}}{T}\,\frac{(1 - \mathrm{e}^{-t_{\mathrm{r}}/T})}{V_{\mathrm{m}}/T}$$

$$= 1 - \mathrm{e}^{-t_{\mathrm{r}}/T}$$

$$= \frac{v_{\mathrm{r}}}{V_{\mathrm{m}}}$$

from equation 8.1. This simply states that for good linearity only a small part of the exponential change should be used—not surprisingly!

Simple ramp generator

Figure 8.2 shows a very simple ramp generator. When the input is high, the transistor is on and saturated, and the output is at 0 V (approx.). When the input goes low, the transistor turns off, and the output starts to rise on a time constant CR towards V_{m}. If V_{m} is much larger than the final value of the output, determined by when the input goes positive, then the linearity will not be too bad. The deviation from linearity is $v_{\mathrm{r}}/V_{\mathrm{m}}$.

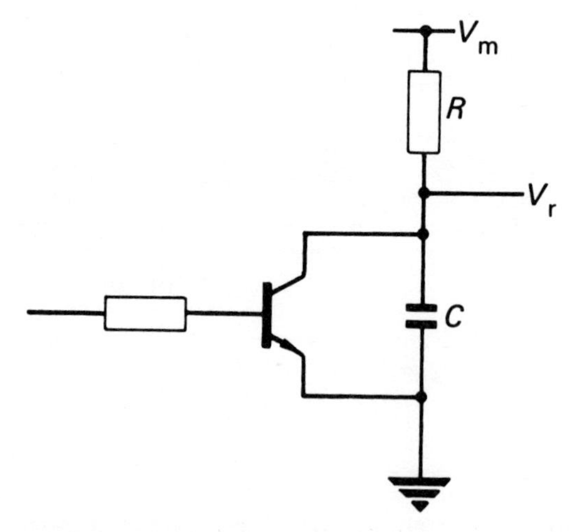

Fig. 8.2 Simple ramp generator

The Miller ramp generator

Figure 8.3 shows the basic elements of a Miller ramp generator. A similar circuit is used in analogue computers as an integrating circuit. In this case it is the constant voltage, V, which is being integrated to give a voltage proportional to time. The triangular device is an amplifier of high gain, such that when the input rises, the output falls. It may be contrasted with the comparator, in that the output is always proportional to the input, and not one of two digital values.

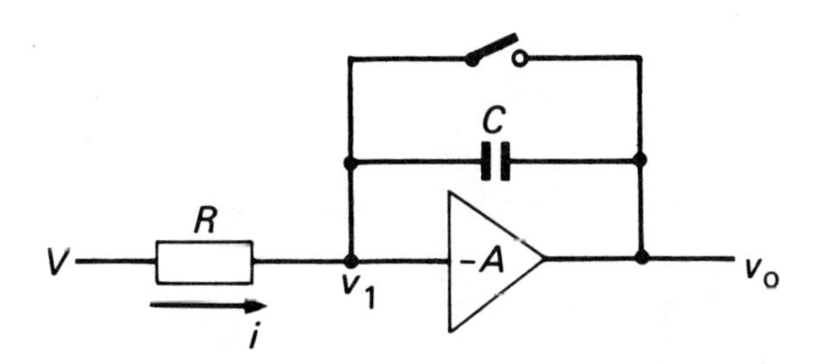

Fig. 8.3 Basic Miller ramp generator

Initially the switch is closed, and hence v_o is equal to v_1. When the switch is opened, a current i flows through R. As the amplifier input impedance is high, the same current flows through C. Hence

$$\frac{V - v_1}{R} = C\frac{d}{dt}(v_1 - v_o)$$

$$v_o = -Av_1$$

$$\frac{V + \frac{v_o}{A}}{R} = -C\frac{d}{dt}\left(1 + \frac{1}{A}\right)v_o$$

$$\frac{dv_o}{AV + v_o} = -\frac{1}{ACR}\left(\frac{A}{1+A}\right)dt$$

$$\ln(AV + v_o) = -\frac{t}{(1+A)CR} + \text{constant}$$

When $t = 0$, $v_o = 0$; hence constant $= \ln AV$.

$$v_o = -AV(1 - e^{-t/(1+A)CR})$$

Comparing this with the previous equation, the circuit is equivalent to a simple arrangement in which the capacitor is $C(1 + A)$, and the resistance is returned to $-AV$ volts. The slope is

$$\frac{V_m}{T} = \frac{AV}{(1+A)CR} \simeq \frac{V}{CR}$$

and the deviation from linearity is V_r/AV. Notice that, although the 'aiming' voltage of the exponential change may be very large ($-AV$), this voltage does not actually exist in the circuit.

Figure 8.4 shows an implementation of the switch of Fig. 8.3. The voltage V is now negative. The input of the amplifier is at about earth potential. When the transistor is turned on it saturates, and the voltage across the capacitor is held at about 1 V. If the base of the transistor is now taken to 0 V, the emitter falls to a negative value, and the

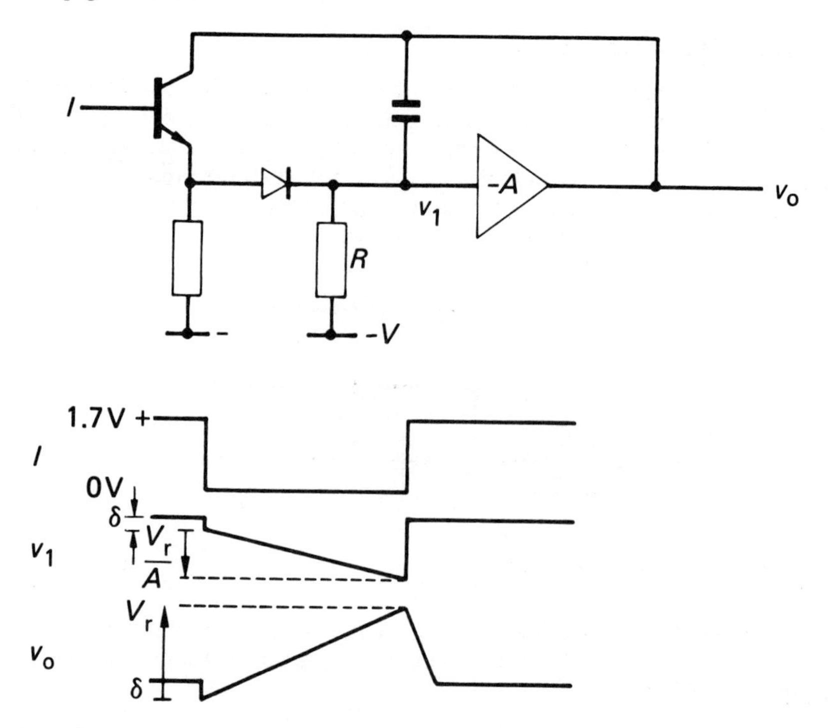

Fig. 8.4 Miller circuit showing implementation of the switch

diode turns off. The amplifier input begins to fall, and the fall is amplified, appearing as a large rise at the output. This, in turn, tends to pull the input positive, thus reducing the extent of that fall. As the input of the amplifier moves only a little (v_o/A), it is often referred to as a **virtual earth**. The output will continue to rise until either the amplifier reaches the limit of its range, or the transistor turns back on. Because the transistor switch can never quite match to the amplifier biasing system, there is always a small step in the output waveform of this type of ramp generator. In some applications (e.g. an oscilloscope time base) this can be ignored, as the beam can be held invisible at this time. In other applications (e.g. digital voltmeters), the step is not acceptable.

The reset time is determined by the current available in the transistor. The amplifier input is held by the transistor emitter. If the available collector current is I, then the discharge time is CV_r/I. This will, of course, be dependent on β.

The resistance R is often chosen to be quite large. The input impedance of the amplifier may not be much larger. In this case the input impedance of the amplifier is in parallel with R, since both have one end to a voltage source. Further, the two resistances act as a potential divider for V, so that the equivalent value for V reduces to $VR_{in}/(R+R_{in})$. For example, suppose $A = 1000$, $V = 25$ V, $v = 20$ V, and $R_{in} = R$. In the ideal case the deviation from linearity is $V_r/AV = 0.08\%$. In the real case, V is reduced to $V/2$, and hence the deviation from linearity is 0.16%.

Bootstrap ramp generator

Consider the simple circuit of Fig. 8.5. When the switch is closed, the capacitor is discharged, and a current V/R flows in the resistor. When the switch opens, this current begins to charge the capacitor. The output of the amplifier rises in sympathy, and so does the voltage V. If the change in V is the same as the change in v_o, then the

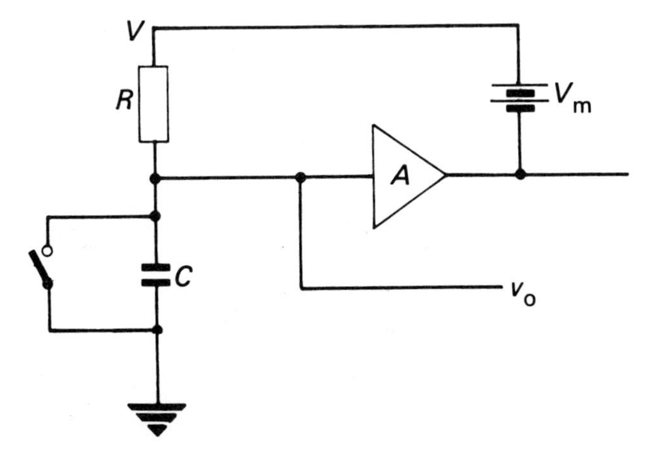

Fig. 8.5 Basic bootstrap circuit

current in R will remain constant, and hence the rate of rise of v_o will also be constant. For V to rise by the same amount as v_o, the gain of the amplifier must be one.

The problem with this circuit is the use of a battery. It can be replaced by a large capacitor such that the charge supplied to the 'ramp' capacitor during the ramp is small compared to that on the 'battery' capacitor in the first place.

Figure 8.6 shows a practical circuit. The amplifier is an emitter follower, whose gain will be close to unity. When its base rises, its emitter also rises, and so does the point A. When A rises, the diode cuts off, and the current in R is supplied by the capacitor C'. The base current of the emitter follower must be kept small compared to that charging the capacitor C, or it must be held constant. As the input impedance of an emitter follower is βR_E, this is not difficult to achieve, provided that R is not too large. The advantage of this circuit is that there is no step in the output at the start of the ramp.

The total charge supplied to capacitor C during the ramp is Cv_r, where v_r is the ramp voltage. This must be supplied by C'. Hence the change of voltage across C', δv is given by

$$C'\delta v = Cv_r$$

The slope of the ramp is given by

$$C\frac{dv}{dt} = i_R = \frac{V - \delta v}{R}$$

$$= \frac{V - Cv_r/C'}{R}$$

$$= \frac{VC' - Cv_r}{C'R}$$

where V is V_p − a diode forward voltage drop.

Rearranging this and integrating such that $v_C = 0$ when $t = 0$, the equation of the ramp is found to be

$$v_C = \frac{VC'}{C}(1 - e^{-t/C'R})$$

From this equation it is noticed that the time constant and the aiming voltage have both been increased by a factor C'/C. The deviation from linearity is v_rC/VC'. If v_r is allowed to reach V (which it may), then the deviation from linearity is simply C/C'.

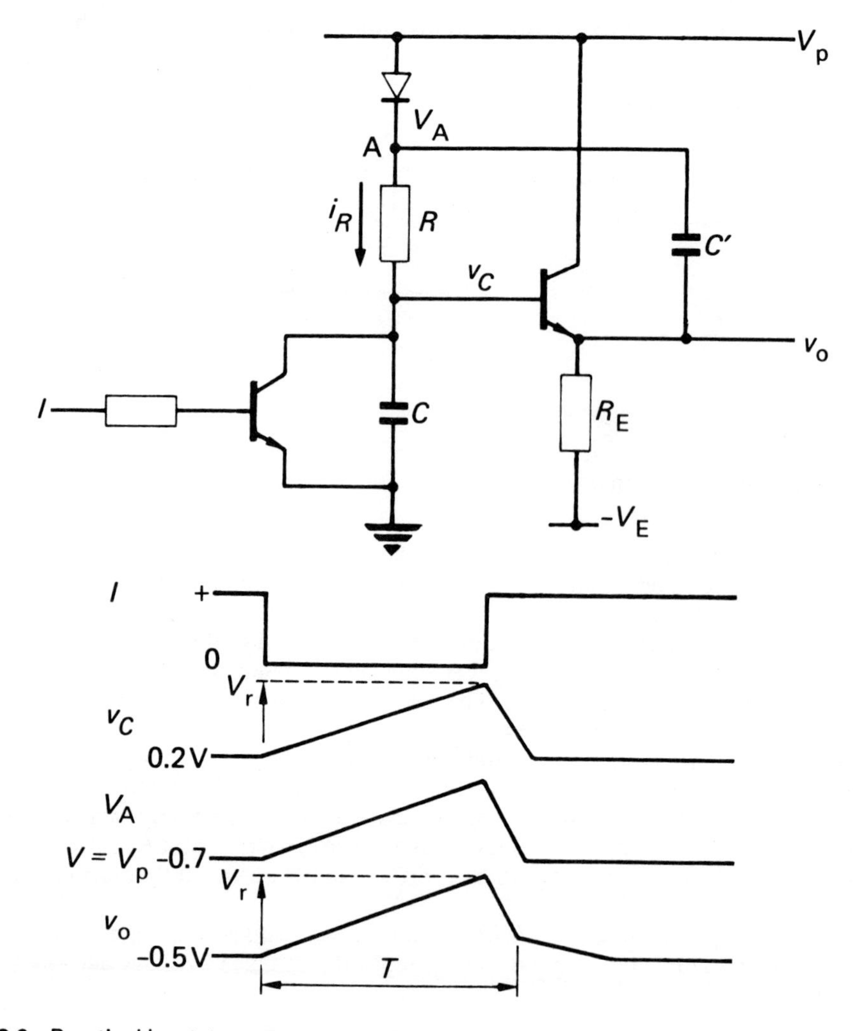

Fig. 8.6 Practical bootstrap ramp generator

The next problem with this circuit is associated with the fact that a real emitter follower has a gain of less than unity.

In this case i_R is given by

$$i_R = \frac{Av_C + V - \delta v - v_C}{R}$$

where A is the gain of the emitter follower.

$$i_R = \frac{1}{R}\left(V - \left(\frac{C}{C'} + 1 - A\right)v_C\right)$$

$$= \frac{VC' - v_C(C + (1 - A)\,C')}{RC'}$$

Thus C in the previous equations is replaced by $C + (1 - A)C'$. The deviation from linearity is given by

$$\left(\frac{C}{C'} + 1 - A\right)\frac{v_r}{V}$$

Note that if $C' = 100C$, and $A = 0.99$, then the gain of the amplifier causes the deviation from linearity to be doubled from what it would have been if A were unity.

Reset time

When the discharge transistor turns on, then βi_b is available to discharge C. However, the voltage across R is reduced below the original value V. Thus the diode will turn on before C is fully discharged. In turn the emitter of the emitter follower is clamped via C', even though the base continues to fall. There is then a second phase to the recovery time while the emitter of the emitter follower falls to allow that transistor to turn on again.

Suppose that the total time from the start of the ramp to the time the emitter follower turns off is T (Fig. 8.7). The charge lost by C' is $C'\delta v = iT$. Hence $\delta v = VT/RC'$. δv will be small, and hence the fall will be almost linear. The total possible fall is approximately V_E. Hence

$$\frac{V_E}{R_E C'} = \frac{\delta v}{t_r} = \frac{VT}{RC' t_r}$$

and hence

$$t_r = \frac{V}{V_E} T \frac{R_E}{R}$$

Now if V is of the same order of magnitude as V_E, and R is of similar magnitude to R_E, then t_r is of the same size as T!

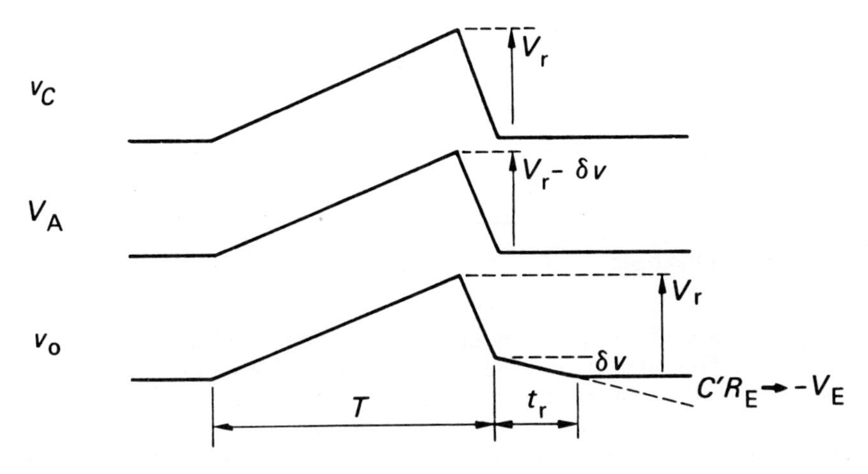

Fig. 8.7 Reset time

Time bases

A time base is most commonly used to deflect the spot of a cathode ray tube across the screen at a constant rate. In a tube with an electrostatic deflection system, this means applying a voltage which varies linearly with time to the deflection plates. The deflection system of a television tube is usually electromagnetic, and is more complex, due to the non-linear characteristics of the deflection coils.

The ramp is controlled by the transistor switch, which discharges the timing capacitor. A Schmitt trigger detects the highest required point on the ramp and switches the transistor on, starting the reset period. In connection with this application, the reset period is often called the **flyback period**, since it is during this time that the spot on the

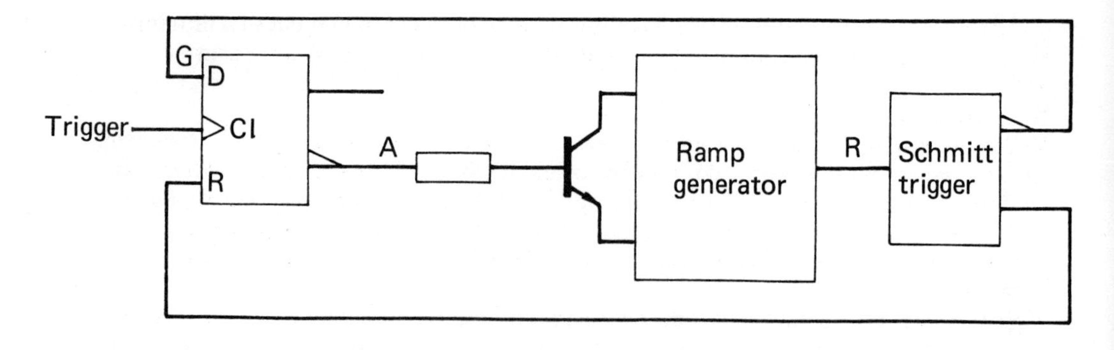

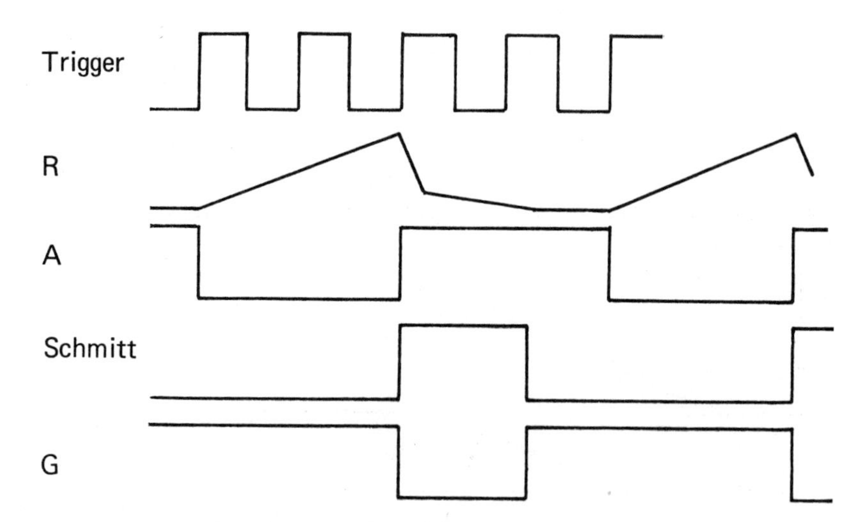

Fig. 8.8 Triggered time base

cathode ray tube screen flies back from the right hand side to the left, ready for the next trace. The end of this period is detected by the other threshold of the Schmitt trigger, which then allows the next ramp to begin. This is thus a free-running time base.

To get a steady picture from a repetitive waveform, it is necessary to delay the start of the next ramp until a given point on the input wave. This point is selected by a trigger circuit, and some logic is then used to control the ramp. Figure 8.8 shows a simple system that should achieve most of the requirements.

In Fig. 8.8 the waveform at A is drawn to correspond with the requirements of turning the discharge transistor on and off. The output of the Schmitt trigger switches when the ramp reaches its maximum point, and switches back when the ramp generator output has reached a point where a new ramp can be allowed to begin.

The problem to be solved now has several facets. The 'first' trigger starts the ramp. Subsequent triggers (of which there may be many if the oscilloscope is displaying several cycles of the trigger waveform) must be ignored until after the flyback period is finished. Finally, if there is a trigger half over when the flyback period ends, the new ramp must not begin until the start of the next trigger pulse occurs. All these cases are illustrated in Fig. 8.8.

The inverse of the Schmitt trigger output is connected to the D input of a flip-flop (positive edge triggered). The 'first' trigger sets the flip-flop, allowing the ramp to begin. The Schmitt trigger output does not change at this time, so the next trigger sets the flip-flop again—which has no effect.

When the Schmitt trigger has changed state, the output causes the flip-flop to reset. The D input now goes to zero, so the next trigger pulse will reset the flip-flop (which it is already), and the discharge can continue. When the Schmitt trigger output returns to the quiescent state, the reset of the flip-flop is removed, and the D input goes to a one. If the input trigger is set at this time, it has no effect, since the flip-flop is edge triggered. Thus the ramp will not restart until the next positive going edge of the input trigger. The D input of the flip-flop cannot be allowed to go to zero until after the end of the ramp, or a trigger pulse occurring during the ramp would cause A to stop the ramp by turning on the discharge transistor.

Digital voltmeter

Another application of a ramp generator is in analogue to digital converters. Figure 8.9 shows a simple system. The analogue input is marked IN. The counter is initially set to zero. The gate output is at zero because the start signal is at zero. The ramp output is low, so the comparator output is at a one.

When the start signal goes to one, the counter begins to count, and the ramp starts. The gate output will follow the clock, and will go to zero when the ramp voltage is greater than the analogue voltage, IN. If the ramp is linear, then the count is a measure of IN. Thus, the analogue signal has been converted to a digital number. With suitable calibration the counter will read the voltage directly, and a digital voltmeter results.

This is by no means the fastest way of performing analogue to digital conversion. The accuracy also relies on the accuracy and stability of the clock. A variation on this theme, however, overcomes most of the accuracy objections, and is used quite often.

Figure 8.10 shows the principle of a dual slope digital voltmeter. The instrument being considered is a 4-digit machine, so that a count of 10 000 would read zero. The counter is set initially to 9000. The switch is set to A in order to sample the signal to be measured. A voltage controlled current generator gives a current proportional to the

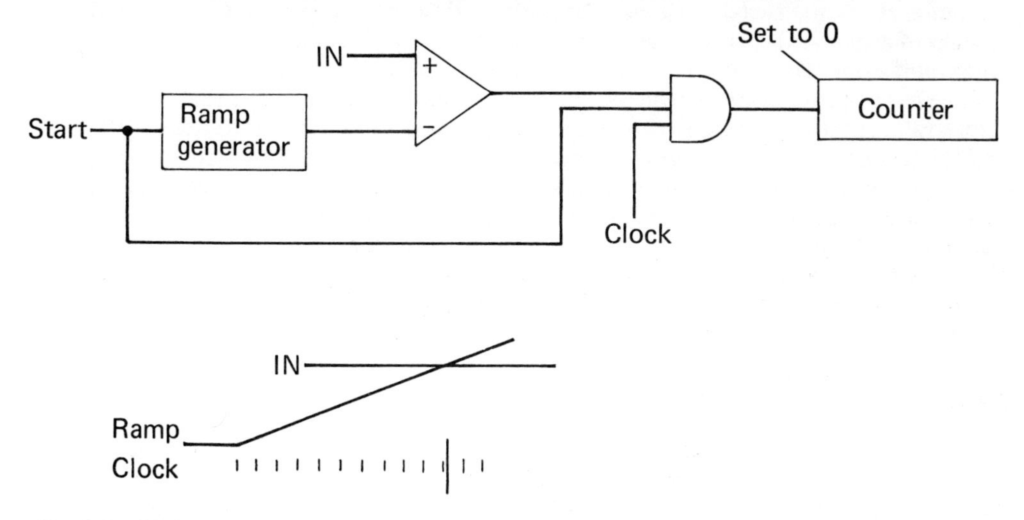

Fig. 8.9 Digital to analogue converter

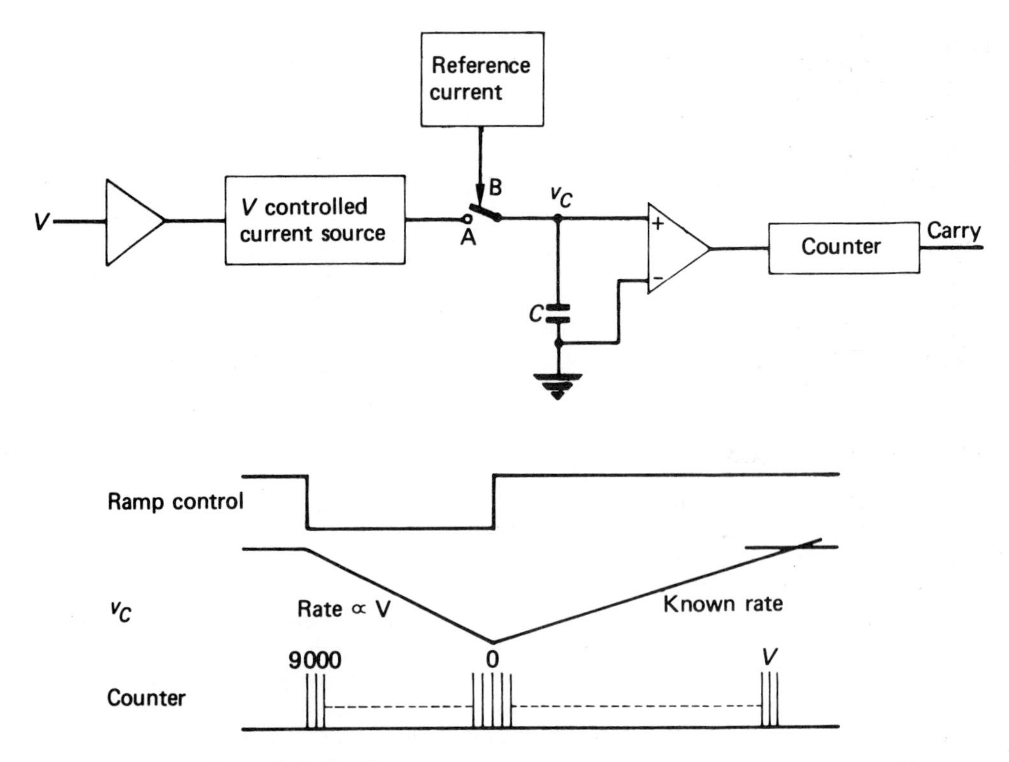

Fig. 8.10 Dual slope digital voltmeter

input voltage into the capacitor, generating a ramp. When the counter reaches 10 000 (reading zero), the switch is changed to position B, and the capacitor is discharged by means of a standard current source. When the input to the comparator reaches zero the count stops. The meter is calibrated to read the voltage directly.

The use of the dual slope principle means that the reading is independent of the clock rate. If the clock runs slowly, then it takes longer for the count to reach 10 000, and hence the capacitor charges to a higher voltage. It will therefore take longer to discharge. But the clock is running slowly , and so the count will be less by a corresponding amount. The two errors compensate precisely, provided that the clock is stable over the short period (less than 1 second) of the measurement, which it should be. A further advantage arises if the sampling of the input is exactly an integral number of cycles of the mains period (multiples of 20 ms in the UK). In this case any 'mains hum' will be averaged out.

Conclusions

A signal whose magnitude varies linearly with time can be used in a number of ways. This chapter describes two ways of generating such waveforms, and outlines two important applications. Very much more needs to be said, but this text is intended to provide only a brief introduction.

Questions

1 Find the initial slope and the deviation from linearity of the ramp generated by the circuit of Fig. 8.2 when $R = 10\ \text{k}\Omega$, $C = 1000$ pF, $V = 10$ V and an input pulse goes from 10 V to 0 V for a period of 1 μs.

2 In the bootstrap circuit of Fig. 8.6, $V_p = 10.7$ V, $R = R_E = 10\ \text{k}\Omega$, $C = 1000$ pF, $C' = 10\,000$ pF and the input goes to 0 V at $t = 0$. Draw the waveforms at A, v_o and v_C. Calculate the voltage of these points when $t = 1$ μs. Assume that the emitter follower voltage gain is 1 and its base current is negligible. The base-emitter voltage is 0.7 V and the collector saturation voltage is 0 V.

What is the deviation from linearity at 1 μs, and what is the flyback time required before a new ramp can be started?

3 What is meant by the deviation from linearity of a ramp waveform?

Sketch the circuit of a ramp generator. Describe its operation, including arrangements for zero-setting. Determine the deviation from linearity of the ramp it produces.

Draw the block diagram of a system for producing the horizontal waveform for an oscilloscope, briefly explaining the function and method of operation of each block in your diagram. Include facilities for triggering from the signal being displayed on the oscilloscope. Show how your system handles potential trigger signals during the ramp and flyback periods. U of M (June 1982)

9 Systems timing

Introduction

There are a number of approaches to the timing of a digital machine. It is also necessary, on occasion, to connect together two or more machines, with different timing mechanisms. This chapter discusses some of the issues involved, and describes some additional circuits that are needed.

Synchronous and asynchronous control

There are two main global approaches to the control of a machine, synchronous and asynchronous. With **synchronous** control, all operations in the machine are **timed** by or **synchronised** to a common **clock**. Generally speaking, information is loaded to a register by the clock signal (Fig. 9.1), and that is followed by a piece of combinational logic. The output of this logic is loaded to another register (or to the same one if it is of suitable design) by the next clock. Provided that there are no undesirable interactions, several operations may go on in parallel. All operations have a defined ending in time, and hence cannot interfere.

With **asynchronous** timing, each module in the machine runs at its own speed, and carries on with a second operation immediately on finishing a previous one, regardless of other modules. Problems arise when two modules wish to communicate. Somehow their timing must be brought into line, or **synchronised**. This can be quite a difficult and time consuming procedure.

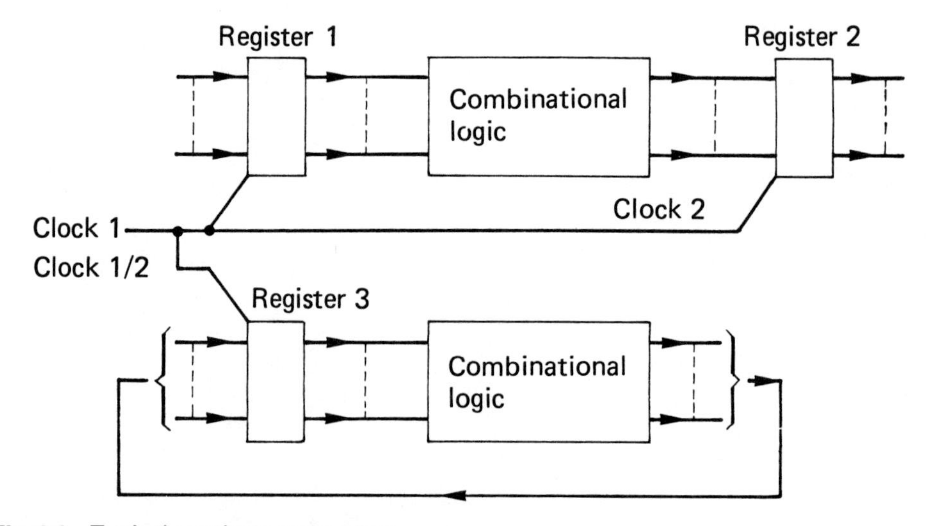

Fig. 9.1 Typical synchronous system

The advantage of using synchronous timing is in the ease of design, commissioning, and maintenance. It is much easier to think only in terms of discrete time slots. On the other hand the system, like a convoy, must run at the speed of the slowest module. Thus asynchronous timing is generally faster.

One disadvantage of 'synchronous' timing is that no system can be completely synchronous. Thinking in terms of even the simple microcomputer, there is no way in which the key strokes of the user's fingers on the keyboard, operating over times of a second or so, can be made in synchronism with the internal CPU clock running on a cycle time of 250 ns.

Thus, every system inevitably has some area where two asynchronous devices have to be synchronised. This is particularly true of distributed systems, where two machines and the transmission medium may all have to be synchronised together.

Races and hazards

Consider the circuit shown in Fig. 9.2. The logic is that of an exclusive-OR, or not-equivalence gate. Both inputs are initially at zero, and both go to one at exactly the same time. The signals A′ and B′ are slightly later than A and B, as shown. This means that for a very short time all the inputs to the AND–OR combination are at one, and the output will change. Once A′ and B′ alter, however, this change will be cancelled.

The reader may experiment with variations on the relative timing of A and B to see how the pulse at C will vary. One can also experiment with an integrated not-equivalence circuit such as the 74LS85, and real pulses such as that shown for C will be observed.

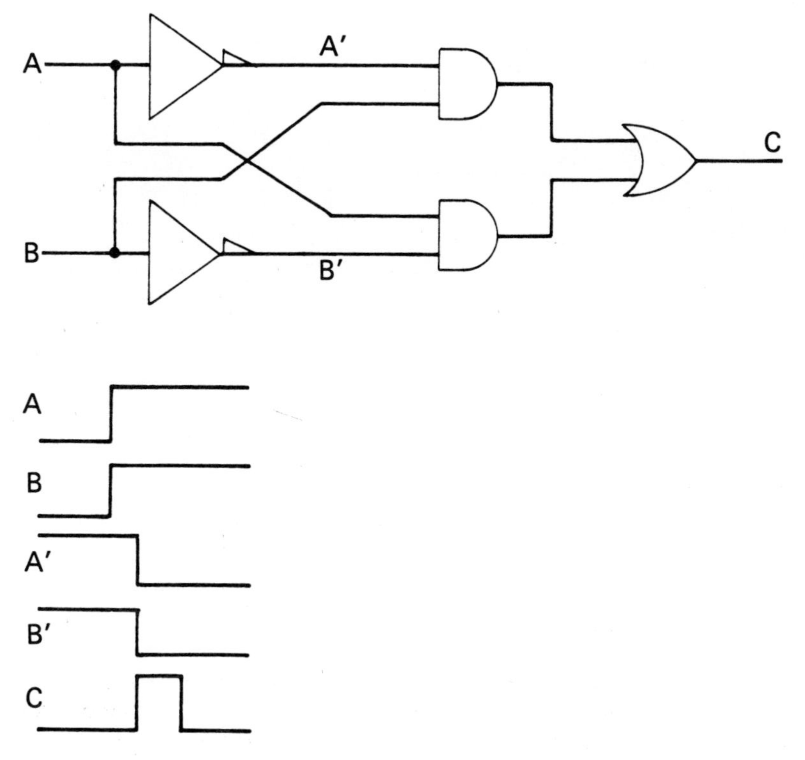

Fig. 9.2 Not-equivalence gate

There are many places in combinational logic where such pulses may occur. Because no two circuits, even of the same type, will have the same time delay, the spikes cannot be eliminated. They are due to differing path lengths through the logic. A properly designed system is such that these unwanted signals do not matter, since they do no harm. One way of ensuring this is to arrange for registers to be clocked only after the longest path is complete. The clock generator circuits must contain no such spikes, possibly being very simple.

A similar problem arises in sequential circuits, where the logic following a register feeds results back to the same register (e.g. the lower half of Fig. 9.1). If the register is constructed from latches, then the clock simply acts as a gating signal and, if the minimum delay through the logic is less than the clock pulse length, there will be an open loop where an output changes an input which changes an output.... The register must therefore be an edge triggered flip-flop, so that changes on the input are locked out as the active clock change takes place.

Set-up and hold times of flip-flops

Even though an edge triggered flip-flop is used, there is still need for care in timing. Consider, for example, the circuit shown in Fig. 9.3, which might represent a counter circuit. The logic is some combinational circuit before the flip-flop. The clock, however, goes directly to the flip-flop.

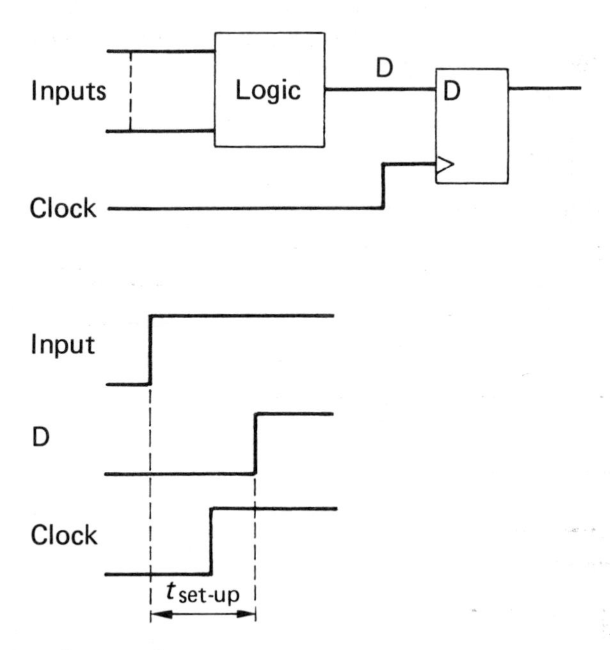

Fig. 9.3 Illustration of set-up time

Suppose that the data changes a very short time *before* the clock, as shown in Fig. 9.3. This change is on the input pins. It takes a measurable time for the results of this change to reach the real input of the flip-flops. The data actually arrives at that point *after* the clock, and hence it is the 'old' data which will be strobed into the flip-flop. To ensure that the data reaches the flip-flop before the clock, a **set-up** time is defined. This will give the time that the data must be steady *before* the clock change to ensure correct operation.

Consider, now, the situation illustrated in Fig. 9.4. The clock input to the IC is buffered in order to drive several flip-flops, only one of which is shown. The data input goes directly to the flip-flop. Because of the delay in the buffer circuit, the data must be held for some time after the clock signal change for the desired operation to be safe. The time that it must be held is called the **hold time**.

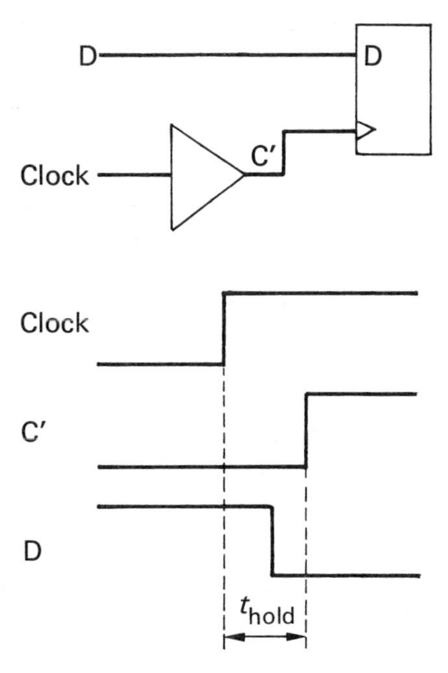

Fig. 9.4 Illustration of hold time

Notice that the circuit of Fig. 9.3 exhibits a positive set-up time, but a *negative* hold time, and the circuit of Fig. 9.4 has a positive hold time and a *negative* set-up time. In practice even a relatively simple circuit such as a flip-flop will have set-up and hold times. These give the conditions under which the device will work to specification. The next section deals with the situation in which the flip-flop settling times may become very long indeed. To understand exactly how these long times arise, the flip-flop circuit has to be examined in great detail.

Synchronisation

It was indicated earlier that, however a machine may be built, there are always occasions when two asynchronous signals will have to be synchronised together. It follows that when this is attempted, there will be occasions when the set-up and hold time rules of the flip-flops will be violated. It is therefore necessary to enquire what will happen under these circumstances, and how adverse effects can be avoided.

Figure 9.5 illustrates what happens if an input signal is gated with the clock signal. IN(a) has a change while the clock is true, resulting in a short pulse in OUT(a). OUT(b) shows a similar effect at the end of IN(b). If these short pulses are used as the clock input to a flip-flop, then they may not be long enough to cause correct operation. Clearly, it is necessary to arrange for the input changes to occur when the clock is low (Fig. 9.5).

Figure 9.6 shows the solution to this particular problem. The signal IN(a) is set to

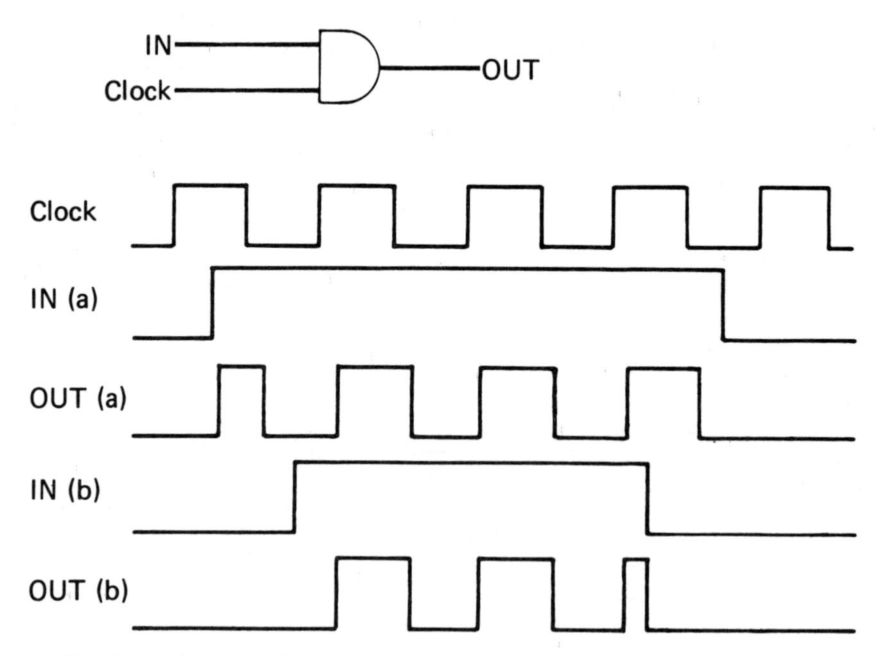

Fig. 9.5 Synchronisation problem

the Q of the flip-flop on the negative going edge of the clock. Thus the clock input to the AND gate can be guaranteed to be zero by the time Q has changed, due to the inherent delays in the circuits. Similarly, the effect of IN(a) going to zero is not communicated to Q until the end of the clock 'one' period. All output pulses are of full length therefore. However, the number of output pulses may be less than at first expected, as illustrated by OUT(b). IN(b) goes to one after the negative going clock edge, so the next 'one' of the clock pulse train will be missed, Q not being changed until the next negative going clock edge.

In spite of the apparent solution to the problem, it is still possible for the input and clock signals to arrive at the flip-flop inputs at times which violate the set-up and hold time rules. Figure 9.7 shows the two inputs happening together. Suppose the time of the change of D to vary from just before to just after the clock change. The output of the flip-flop will be one of the curves illustrated in a single drawing at Q.

If this is the output of a latch which is frozen when the clock goes high (as opposed to an edge triggered flip-flop), and is used as the input to a second latch which uses the opposite phase of the clock as its clock signal, then Q2 shows two of the possible results. (This is the internal operation of an edge triggered flip-flop.) In any case, the time for Q to reach a final state may be very much longer than the time specified in the data sheet on the assumption that set-up and hold times rules were obeyed.

There have been many attempts to overcome this difficulty. While the effects can be reduced by good design, they can *never* be completely eliminated. A certain amount of energy is needed to cause the flip-flop to switch. The amount of energy required cannot be reduced too much, or any noise (unwanted signal) picked up from adjacent circuits will cause random switching. Thus it will always be possible to supply exactly half the energy required to cause switching, and hence the flip-flop will only half switch. If the energy supplied is just less than or just more than this, then the switching will be slow until any positive feedback can take effect. Detailed analysis of the situation is beyond the scope of this text. The result of an analysis by Kinniment *et al* (see Appendix A) shows that for reliable operation of a system one must wait for some

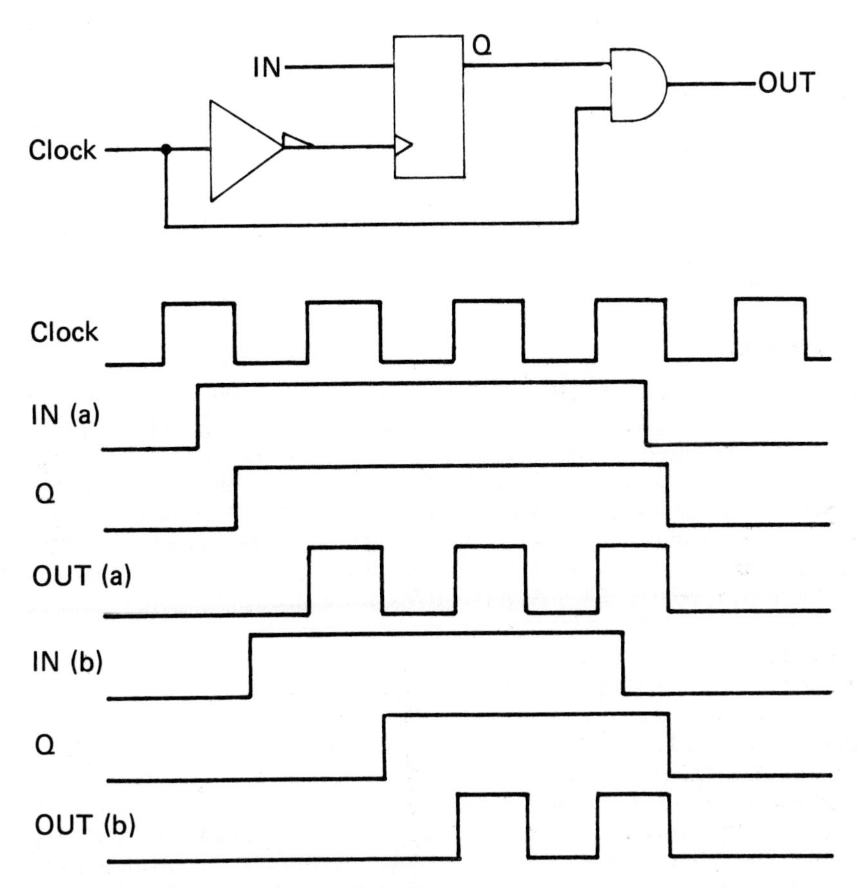

Fig. 9.6 Synchronisation of data to a clock

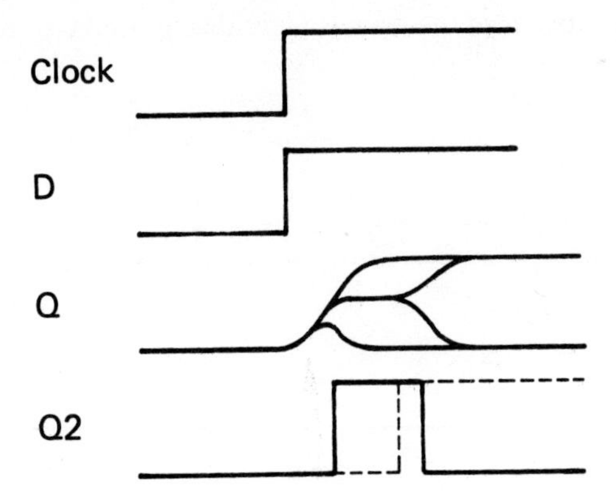

Fig. 9.7 Set-up and hold time rules disobeyed

time after the clock pulse for the output of the flip-flop to settle. Table 9.1 shows some results for flip-flops built with two different technologies being clocked at a rate of 10 MHz. The first two columns show the waiting time after the clock edge, and the last column gives the mean time between occasions when the wait was too short.

Table 9.1 Settling times of flip-flops (ns)

TTL (N)	ECL 10K	MTBF
41	18	1 hour
46	21	1 day
49	23	1 week
57	27	1 year
60	28	10 years

Clock skew

A serious problem can arise in the clocking of registers, and this problem gets worse as circuit speeds rise. It is due to the clock signal taking a finite time to travel from the driver to the chips where it is used. The time to different chips within the register will be different, and hence the data will be loaded at noticeably different times. A similar problem occurs when a signal is used to gate a set of data onto a parallel highway or bus. The difference in time between earliest and latest clocks, measured at the flip-flop input, is known as the **skew**. Figure 9.8 shows a clock driver physically located such that the skew will be large. Clearly, if the driver was placed at the arrow, the skew would be halved.

The effects of skew must be considered carefully, and allowance made for the difference of timing of the flip-flops at B and C in Fig. 9.8. The effects become more important as the requirement for speed increases. Indeed, the clock skew can cause very severe loss of speed. The only way to reduce the effects is to take great care with the physical layout of the circuit. In fact minimisation of skew is not always essential. If all registers in a system are distributed in the same way, and if the longest connection between flip-flops is over, say, two or three stages, then the system of Fig. 9.8 is quite acceptable. The clock arrives late at C, but it also arrives late at equivalent stages in subsequent registers in the system. In one system known to the author, a clock pulse of 10 ns was used. The clock at one end of the register set was measured to be over 5 ns later than the clock at the other end. But the system ran successfully with a clock period of 20 ns, simply because the signals were distributed correctly.

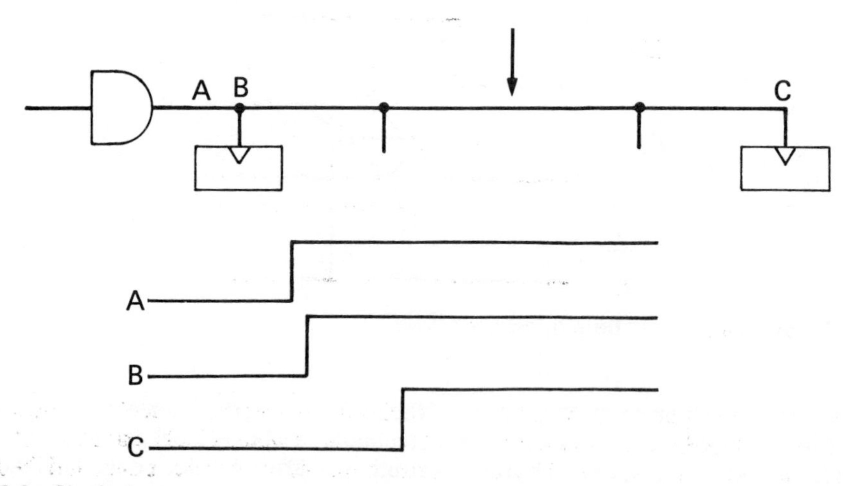

Fig. 9.8 Clock skew

Phase locked loops

There are many occasions when a system must be synchronised to an external bit stream, rather than synchronising the external signal to the machine. For example, the stream of data coming off a magnetic disk system often contains the timing information within it. The precise timing depends on the rotation speed of the disk, which, in turn, is dependent on the mains frequency. The mains frequency can vary by ±2%. On a 5 inch diameter track on a floppy disk there may be up to 90 000 bits of information. A 2% error represents about 2000 bits! Thus the data can only be understood if the inherent timing information is used for synchronisation. Another case is where the data is arriving via a telephone line, so that the clock at the transmitting end cannot by synchronised exactly with that at the receiver.

To achieve the required synchronisation, a phase locked loop is used.

Figure 9.9 is a block diagram of a phase locked loop. The phase detector compares the incoming signal with that from a voltage controlled oscillator. If they are not exactly in synchronism, a difference signal appears. After filtering, a steady voltage proportional to the difference is produced. This, in turn, is used to control the oscillator frequency in such a way as to reduce the difference (see Chapter 6 for the design of voltage controlled oscillators).

In a digital system, a not-equivalence gate can be used as the phase detector. Figure 9.10 illustrates the principle. Suppose, first, that the input and output are exactly in synchronism (a). As the inputs are always equivalent, a not-equivalence gate will always give zero output. Consider, next, case (b). The output of the detector is a pulse waveform whose mark–space ratio varies with the difference between the two input signals. If the two signals are completely mismatched, then the output is always a one.

If the not-equivalence gate output is now fed to a circuit which determines the average of its input, and maintains it at a steady voltage level, then this average can be used to control the frequency of the voltage controlled oscillator. Such a circuit is one application of a device known as a **low pass filter**. It must maintain the frequency of the voltage controlled oscillator long enough for the system to remain in synchronism over a long period of time compared to the bit time—perhaps 10 bit times.

The above method of designing a phase locked loop makes use of some analogue techniques in the low pass filter. Figure 9.11 shows the principle of a completely digital phase locked loop.

An oscillator running at a high frequency is used to clock a counter. The output of the counter will be at the expected input frequency. The frequency division should be as great as is necessary to achieve the required control. The input signal is sampled by the counter output. If the sample is a zero, then the next clock to the counter is inhibited. If the sample is a one, then an extra clock is fed to the counter. In Fig. 9.11, Counter A is slightly displaced from the input. At sample time a one is detected, and

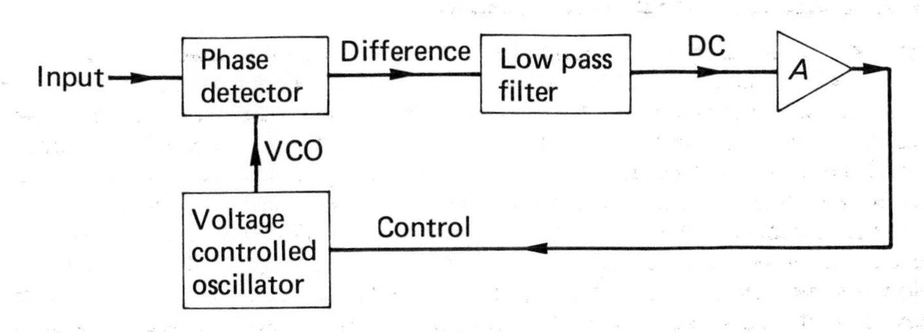

Fig. 9.9 Phase locked loop

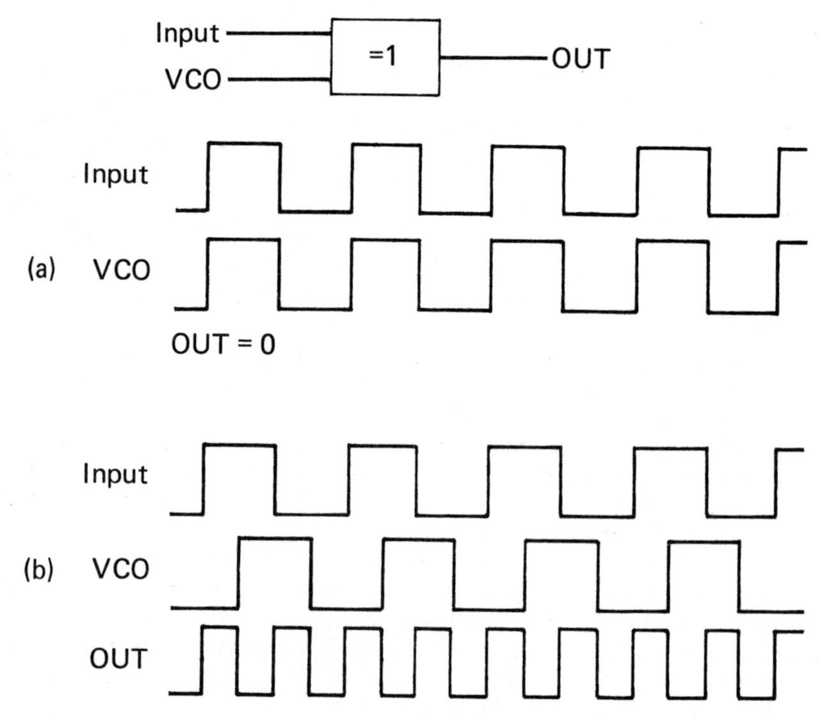

Fig. 9.10 Not-equivalence method of phase detection

the counter receives an extra clock. The next sample will therefore be taken a little earlier relative to the input, assumed to occur at a stable clock period.

Counter B is where the counter occurs displaced in the opposite direction to that of Counter A. In this case a clock to the counter is removed, so that the next sample will be a little later with reference to the input.

The insertion and removal of clock pulses to the counter may be done via a multiplexer on the input of the counter. It is quite a tricky business making sure that the extra clock does not occur too soon after a 'regular' one (so that set-up and hold times are obeyed), and that only one clock pulse is removed. The reader may find it a useful exercise to try (and don't forget the effects of not observing set-up and hold times!).

Synchronisation of remote systems

In any data stream there will be periods when there are few logic changes (transitions). A 'self timed' data code is designed such that there is a maximum time between transitions—say every 10 bits in the standard asynchronous codes. Figure 9.12 shows a typical data stream. If this is differentiated, a spiky waveform results, which is then rectified. The new data stream contains a frequency component at the clock frequency. If the low pass filter is sufficiently slow in its response, then the oscillator frequency will be held constant. In the case of the fully digital filter, the data transitions will determine when pulses are added to or subtracted from the input to the counter. As the low pass filter must have a slow response, it will be necessary for a new data stream to have a 'leader' of signals designed to allow the receiver to achieve synchronisation. Provision of such a leader is common practice.

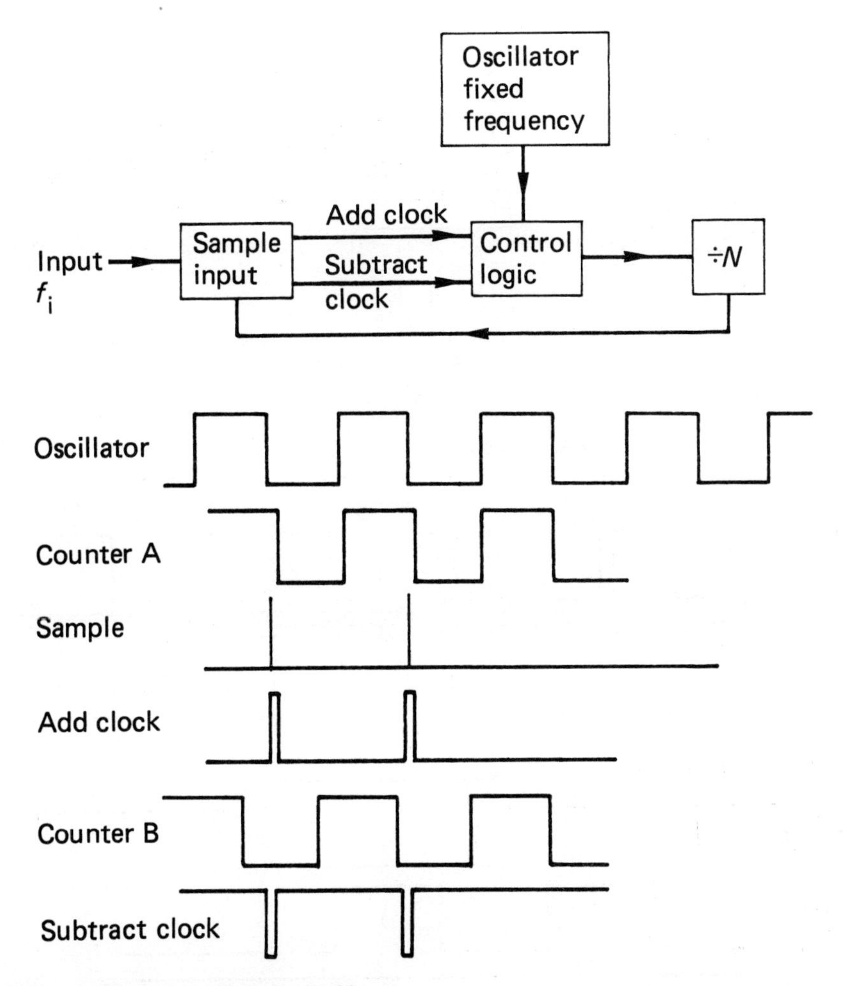

Fig. 9.11 Fully digital phase locked loop

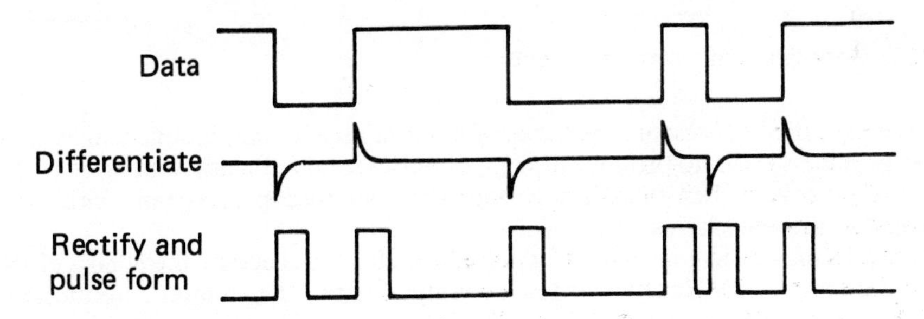

Fig. 9.12 Bit synchronisation

Multiphase clocks

Many early computers made use of repetitive trains of pulses to control their timing. Modern microprocessors are again doing something similar, though the number of pulses in a train is usually much less. At the opposite end of the scale, it is a requirement

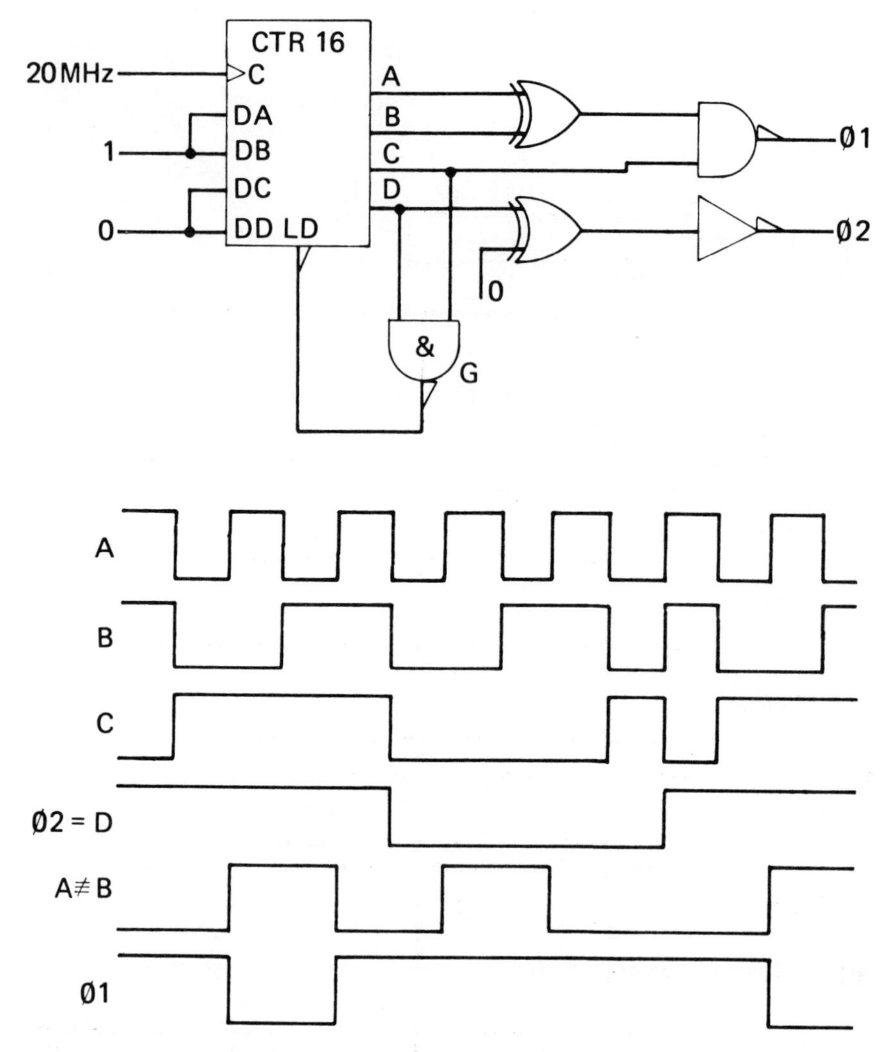

Fig. 9.13 Generation of a two phase clock

that large pipelined processors should be able to sequence the pipeline stages correctly. Two problems arise: to produce a string of pulses, and to ensure that none of these pulses overlap, even when considering times at the microscopic level and with very slow edges to the waveforms.

Figure 9.13 shows the generator of a two phase clock, as used by the 8080A eight-bit microprocessor. A 20 MHz clock drives a counter circuit. The counter runs until the state 1100 is detected by the NAND gate, G. The counter is now put in the load state. The data inputs are wired to be 0011, so the next clock pulse sets the counter to this state, incidentally returning it to the count state. Once the first cycle after switch on is complete, the circuit operates as shown by the waveforms in Fig. 9.13. The clocks to the 8080A are Ø1 and Ø2, and are true when *low*! It will be seen that the low parts of these two waveforms do not overlap.

A multiphase clock can be generated by using a counter in a similar manner, and decoding the outputs with a set of AND gates. If the clocks are to be non-overlapping, the AND gates can include the clock among the inputs. This ensures that an output only appears in the last half of each clock period. The reader may like to derive a four

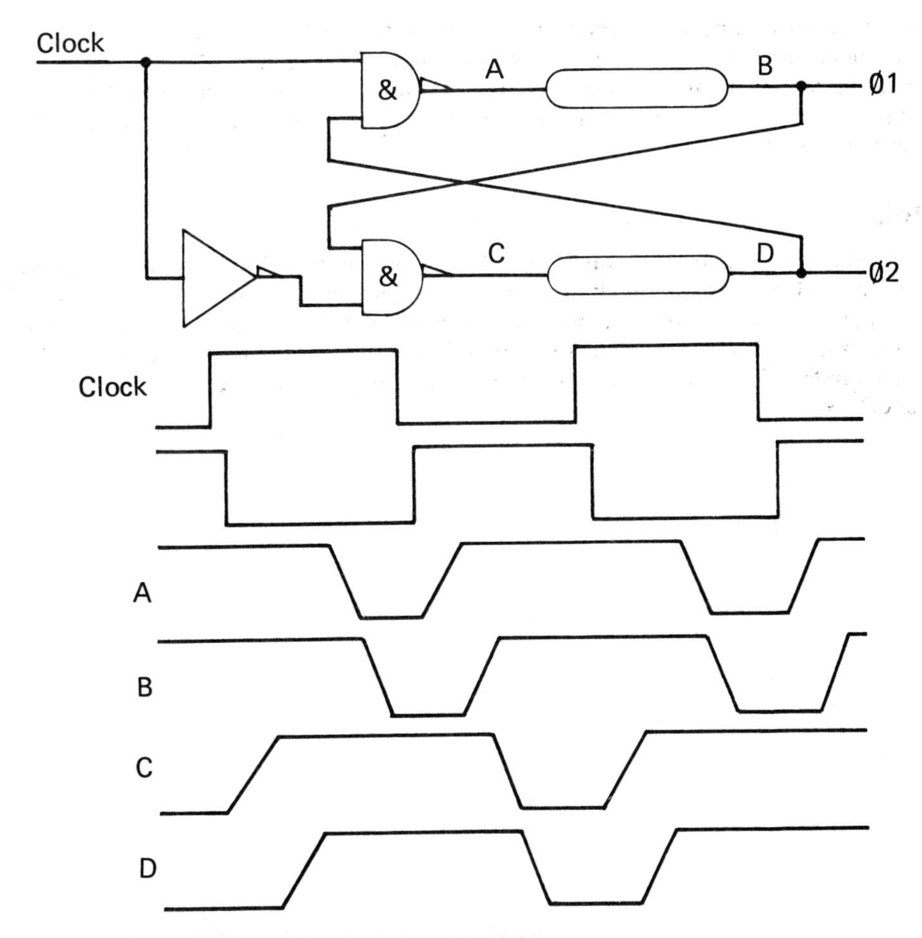

Fig. 9.14 Non-overlapping clocks using gate delays

phase clock using waveforms A, B and C in the first eight beats of waveform A of Fig. 9.13. A is the clock, and B and C can be decoded to give the four states. Three-input gates will be needed.

The 8080A required external clock generation. Most modern microprocessors have on-board clock generation from a single external crystal controlled oscillator. The gate delays on a chip are very much the same for the same gate types, so that the time through a gate can be used as a reliable delay. While the value may not be known in absolute terms, it is certain that if one gate delay on the chip is long, then all the other gate delays on the SAME chip will also be long. Edge times may also have effects.

Figure 9.14 shows an appropriate circuit and its waveforms, and is derived from that of Mead and Conway (see Appendix A). The delay elements would be made up of a pair of inverters designed to increase the drive capability of the clock signal, and therefore relatively slow. The waveforms are straightforward to generate, assuming the switching points to be half way between the logic levels. Again, it is seen that the low portions of B and D do not overlap.

Conclusions

This chapter discusses a number of system problems involving timing. A very large proportion of problems faced in design and commissioning of systems is due to the difficulties of timing, whether due to races and hazards, to clock skew, to synchronisation of asynchronous events, or to extracting data from a bit stream from which the clock information must also be extracted. With the advent of custom VLSI chips, the problem becomes even more acute, since modifications cannot be made after manufacture.

Finis

This text does not claim to cover all possible aspects of the subject. To a large extent it seeks to teach by example. If the reader properly understands the 'how' and 'why' of the examples, then it should be possible to understand other circuits. To design new ones is more difficult—but then creation of the new is always more difficult than analysis of the old. Nevertheless, it is hoped that the basic knowledge required can be found here.

Appendix A
Further reading

AHMED and SPREADBURY, *Electronics for Engineers*, Chapter 5 (Feedback). Cambridge UP (1973)

BOYLESTAD and NASLELSKY, *Electronic Devices and Circuit Theory*, Chapter 16 (A/D and D/A techniques). Prentice Hall International (1982)

CARR, *Digital Interface with an Analogue World*, Chapters 10–12 (A/D and D/A techniques). Tab Books (1978)

GARNER, *Phaselock Techniques*. Wiley (1979)

GIBSON, *Electronic Logic Circuits*, 2nd Edition. Edward Arnold (1983)

KINNIMENT *et al*, Synchronisation and Arbitration Circuits in Digital Systems. *Proc. IEE* **123**, pp. 961–966 (1976)

KINNIMENT *et al*, Circuit Techniques in Large Computer Systems. *Radio and Electronic Engineer* **43**, pp. 435–441 (1973)

MATTICK, *Transmission Lines for Digital and Communications Networks*. McGraw Hill (1969)

MEAD and CONWAY, *Introduction to VLSI Systems*, Chapter 7 (timing on ICs). Addison-Wesley (1980)

OPEN UNIVERSITY, *Electromagnetics and Electronics TS282*.
Unit 6 Transient Response
Unit 7 Sinusoidal Response
Unit 8 Network Modelling and Analysis
Units 14 and 15 Time Base Generator

STONE, *Microcomputer Interfacing*. Addison-Wesley (1982)

Appendix B
Outline solutions to questions

Chapter 2

1 $2.2 \times 200\,\Omega \times 50\,\text{pF} = 22\,\text{ns}$.
2 (a) Ramp of 0.025 V amplitude swinging about $V/2$.
(b) A 'square' wave with 0.025 V drop on top and bottom. Swings about 0 V. 100 kHz.
(c) Half-period of 5 μs; rise and fall time of 2.2 μs.
(d) Spiky waveform with rise and fall time of 2.2μs. 1 kHz.
(e) Square wave—the effect of the capacitor is negligible at this time scale.
(f) Very narrow spikes.
3 11 ns.
4 Rising edge: $2.2 \times 20\,\Omega \times 10\,\text{pF} = 440\,\text{ps}$.
Falling edge: $2.2 \times 1\,\text{k}\Omega \times 10\,\text{pF} = 22\,\text{ns}$.
Diode turns off.
5 $2.2 \times 72\,\text{pF} \times 9/10\,\text{M}\Omega = 143\,\mu\text{s}$.
6 8 pF.
7 Input to oscilloscope rises to 6/78 of input volts instead of 7.8/78. Time constant of 70 μs for the rest of the rise.
8 Input rises to 10/82 of input voltage instead of 8.2/82. Time constant of 74 μs for the fall to final value.

Chapter 3

1 For Fig. 3.3: $3R_1 = R_2$; values of a few kΩ are suitable.
For Fig. 3.4: $4R_3 = R_4$.
2 Input at 4 V, about to switch; input current = 1 mA.
Hence $R_1 = 4\,\text{k}\Omega$.
$1\,\text{mA} + 0.2/R_2 \geqslant 5/R_3$
$1/4\,\text{mA} + 3.5/R_2 \leqslant 5/R_3$
Hence $R_2 = 4.4\,\text{k}\Omega$ and $R_3 = 4.8\,\text{k}\Omega$.

Chapter 4

1 For saturation of the transistor, $R < 4.3 \times 40/5 = 34\,\text{k}\Omega$.
Choose $R = 10\,\text{k}\Omega$
$4 = 8.3(1 - e^{-20/CR})$ giving $CR = 30\,\mu\text{s}$; $C = 0.003\,\mu\text{F}$.
For short input, output is same length: 10 μs.
2 Waveforms—see Fig. 4.3.
Voltage across R given by $(5 - 1.1)\,R/(20 + R) < 0.4$
$R < 2.3\,\text{k}\Omega$.
3 Waveforms—see Fig. 4.5.
$0.45 = 4.35(1 - e^{-t/CR})$ giving $t = 0.107CR$
Pulse $= t + 4\,\text{ns}$

Short T: For recovery time see text: or D never changes.

4 $CR = 1870$ ns; e.g. $R = 470\,\Omega$, $C = 0.004\,\mu$F.
Output resistance of an ECL gate $\simeq 5\,\Omega$;
recovery time $= 3 \times$ time constant $= 60$ ns.

5 See Fig. 4.3 for example.
$1.6\,\text{mA} \times R << 0.4\,\text{V}$; choose $R = 100\,\Omega$.
$(1.6 - 0.36) = (3.5 - 0.36)\,(1 - e^{-t/CR})$ with $R = 220\,\Omega$
$C = 0.0018\,\mu$F.
Recovery time: $(3.5 - 0.8) = (3.5 - 0.36)\,(1 - e^{-t/CR})$ with $R = 110\,\Omega$;
giving recovery time $= 390$ ns
Minimum pulse width; determined by rise times and circuit characteristics.

6 Note that *negative* logic is being used. Calculation is similar to questions 3 and 4.
Output is $t + 2$ ns.

Chapter 5

1 $3\,\text{V}/1\,\text{k}\Omega = 1\,\text{mA} \times 40 = 40\,\text{mA}$.
During fall an average of 5 mA is supplied to the resistor.
Fall time $= 10^{-9} \times 5/35 \times 10^{-3} = 143$ ns (0 – 100%).
Rise time $= 2.2 \times 470\,\Omega \times 10^{-9} = 1\,\mu$s.

2 Threshold may be going up or down—not specified.
Steady state with threshold at 0 V, out low.
See Fig. 5.8 for waveforms. If trigger stays low, output remains high (555 timer specification).
$2V/3 = V(1 - e^{-t/CR_T})$ giving $CR_T = 0.909$.
$C < 10\,\mu$F; $R_T < 20\,\text{M}\Omega$.
Suggest 910 kΩ, 1 μF.
See Fig. 5.3.

3 See, for example, Figs 5.5, 5.7, 5.8 and 5.9.

4 $100\,\text{pF} < C < 10\,\mu\text{F}$; $5\,\text{k}\Omega < R < 6.6\,\text{M}\Omega$.
$t = 1.1\,CR_T$; $CR_T = 182$ ns, 1.82 s.
$R_T = 5\,\text{k}\Omega$, $C = 40$ pF; need some compromise here.
$R_T = 1.8\,\text{M}\Omega$, $C = 1\,\mu$F might be acceptable.
Recovery times: $10\,\text{V} \times 40\,\text{pF}/32\,\text{mA} = 12.5$ ns (about 3 mA in R).
$10\,\text{V} \times 1\,\mu\text{F}/35\,\text{mA} = 285\,\mu$s.

Chapter 6

1 $t = 4\,\mu\text{s} = C(R_1 + 2R_2)\ln 2$ (Fig. 6.1)
$C(R_1 + 2R_2) = 5.8\,\mu$s.
Let $C = 1000$ pF, then $R_1 + 2R_2 = 1\,\text{k}\Omega$; too small.
100 pF 10 kΩ.
e.g. $R_1 = 4.7\,\text{k}\Omega$, $R_2 = 2.7\,\text{k}\Omega$.
Mark–space ratio cannot be 1 : 1.

2 R_2 cannot be too small—say 470 Ω. R_1 can be as low as 1 kΩ say. $C \geqslant 100$ pF.
Hence $0.69 \times 100\,\text{pF} \times 2\,\text{k}\Omega = 138$ ns—about 7 MHz.
$R_1 + R_2 \leqslant 6.6\,\text{M}\Omega$. $C \leqslant 10\,\mu$F. For reasonable mark–space ratio, let
$R_1 = R_2 = 3.3\,\text{M}\Omega$.
Hence $t = 69$ s giving 1/70 Hz.

3 Fig. 6.5.
From previous section of text, period $= 2\,CR \ln 3$.
If $C = 100$ pF, $R = 47\,\Omega$; i.e. parallel match to a 50 Ω line!

4 Let $I = 1$ mA; swing $= 1$ V; $R = 1/2I = 500\ \Omega$.
$I = C\mathrm{d}V/\mathrm{d}t$; giving $C = 250$ pF.
The two ends of the capacitor swing from -1.4 V steady level, to -0.4 V and -2.4 V; see Fig. 6.9.

5 This circuit is analysed in a manner similar to that of Fig. 6.3, with similar results.

$$t_1 = C_1R_1 \ln \frac{3.5 - (y - 3.5)}{3.5 - y} = C_2R_2 \ln \frac{3.5 + x}{x}$$

where y is the point reached by the exponential on A when switching occurs, and x is the point reached by the exponential on B when switching occurs.

$$t_2 = C_1R_1 \ln \frac{3.5 + y}{y} = C_2R_2 \ln \frac{3.5 - (x - 3.5)}{3.5 - x}$$

If $x = y = 1.75$, $C_1R_1 = C_2R_2$ and $t_1 = t_2 = CR \ln 3$, which is independent of the logic levels.

Chapter 7

1 Relative permittivity = 3.

2 Characteristic impedance = 140 Ω.
(a) A goes to 5 V at time 0, and 3.3 V at time 11.6 ns.
B goes to 3.3 V at time 5.8 ns.
(b) A goes to 5 V at time 0; B to 5 V at time 5.8 ns.
(c) A goes to 5 V at time 0, and to 6.7 V at time 11.6 ns.
B goes to 6.7 V at time 5.8 ns.

3 (a) A goes to 5 V at time 0, to 0 V at time 11.6ns, and then rises on a time constant 1.4ns to 10 V.
B rises on a time constant of 1.4 ns to 10 V, starting at time 5.8 ns.
(b) As for part (a), except that the exponential rise is to 5 V instead of 10 V and a time constant of 0.7 ns.

4 A goes to 350 mV at time 0, and 700/3 mV at time 30 ns.
B goes to 700/3 mV at time 15 ns.
C goes to 350 mV at time 7.5 ns, and to 700/3 mV at time 22.5 ns.

5 30.6 ns; 81.6Ω.

6 10 ns; 50 MHz.

7 Delay on printed circuit board approx. 6 in. (150 mm) per ns.
Input of the line rises to 0.6 V at time 0, to 0.696 V at time 2 line delays, and falls to 0.692 at 4 line delays.
Output of the line rises to 0.72 V at 1 line delay, and falls to 0.692 V at 3 line delays.

8 (a) Input of line has a pulse of amplitude $V/2$, and width of 2 line delays.
(b) Input as (a), followed by an exponential rise of time constant CR_0 to V.
Output—an exponential rise from 0 to V on a time constant CR_0, starting at time 1 line delay.
(c) As (b), but the initial pulse at the input is to V and a time constant of $CR_0/2$.
Discuss parallel matching, loadings of C, and effect on R_0; open-collector and high impedance drivers; resistance pull ups.

9 Series and parallel matching.
50 kΩ/4 is very high relative to R_0; ignore.
Series matched line—68 Ω in series at source, 16 pF load; waveforms as for question 8 (b); time constant = 16 pF × 75 Ω = 1.2 ns.
Parallel matched line—75 Ω to AC earth at termination.
Waveforms as question 8 (c); time constant 16 pF × 75 Ω/2 = 0.6 ns.

Split loads into 2 groups in serial case—time constant now as the parallel case.
Series matched line—maximum; delay now 2 line delays. Dangerous.
Parallel matched line—R lowered to 67.9 Ω and delay increased to 5.84 ns.

10 Rise time > 2 × line delay; hence line delay < 1 ns; i.e. 6–8 in (150–200 mm). See question 8 for waveforms.

11 Use method of Fig. 7.13, and hence rise time using Fig. 2.1 (ignore 55 kΩ). At point A in diagram replace 2 lines by a termination of $R_0/2$. Hence calculate $v_o = 2v_f/3$. This voltage *must* propagate down *both* branches.

Chapter 8

1 1 V/μs; 9.5%.

2 A = 10.895 V, $v_o = 0.295$ V, $v_e = 0.995$ V, deviation = 1%, flyback time = 1 μs ($v_E = -10$ V).

3 See text.

Index